Chemical Reaction Kinetics

Chemical Reaction Kinetics

Concepts, Methods and Case Studies

Jorge Ancheyta
Instituto Mexicano del Petróleo
Mexico City, Mexico

Registered Offices
John Wiley & Sons, Inc., 111 River Street, Hoboken, NJ 07030, USA
John Wiley & Sons Ltd, The Atrium, Southern Gate, Chichester, West Sussex, PO19 8SQ, UK
Wiley-VCH Verlag GmbH & Co. KGaA, Boschstr. 12, 69469 Weinheim, Germany
John Wiley & Sons Singapore Pte. Ltd, 1 Fusionopolis Walk, #07-01 Solaris South Tower, Singapore 138628

Editorial Office
9600 Garsington Road, Oxford, OX4 2DQ, UK

For details of our global editorial offices, customer services, and more information about Wiley products visit us at www.wiley.com.

Wiley also publishes its books in a variety of electronic formats and by print-on-demand. Some content that appears in standard print versions of this book may not be available in other formats.

Library of Congress Cataloging-in-Publication Data

Names: Ancheyta, Jorge.
Title: Chemical reaction kinetics : concepts, methods and case studies / Prof. Jorge Ancheyta.
Description: Hoboken, NJ : John Wiley & Sons, Inc., 2017. | Includes bibliographical references and index.
Identifiers: LCCN 2017004766 (print) | LCCN 2017005385 (ebook) | ISBN 9781119226642 (cloth) | ISBN 9781119226659 (Adobe PDF) | ISBN 9781119227007 (ePub)
Subjects: LCSH: Chemical kinetics. | Chemical reactions.
Classification: LCC QD502 .A53 2017 (print) | LCC QD502 (ebook) | DDC 541/.394–dc23
LC record available at https://lccn.loc.gov/2017004766

Cover Design: Wiley
Cover Image: © alexaldo/Gettyimages

Set in 10/12pt Warnock by SPi Global, Pondicherry, India
Printed and bound in Malaysia by Vivar Printing Sdn Bhd
10 9 8 7 6 5 4 3 2 1

Contents

About the Author

Jorge Ancheyta, PhD, graduated with a bachelor's degree in Petrochemical Engineering (1989), master's degree in Chemical Engineering (1993) and master's degree in Administration, Planning and Economics of Hydrocarbons (1997) from the National Polytechnic Institute (IPN) of Mexico. He splits his PhD between the Metropolitan Autonomous University (UAM) of Mexico and the Imperial College London, UK (1998), and was awarded a postdoctoral fellowship in the Laboratory of Catalytic Process Engineering of the CPE-CNRS in Lyon, France (1999). He has also been visiting professor at the Laboratoire de Catalyse et Spectrochimie (LCS), Université de Caen, France (2008, 2009 and 2010), Imperial College London, UK (2009), and Mining University at Saint Petersburg, Russia (2016).

Prof. Ancheyta has worked for the Mexican Institute of Petroleum (IMP) since 1989, and his present position is Manager of Products for the Transformation of Crude Oil. He has also worked as professor at the undergraduate and postgraduate levels for the School of Chemical Engineering and Extractive Industries at the National Polytechnic Institute of Mexico (ESIQIE-IPN) since 1992 and for the IMP postgrade since 2003. He has been supervisor of more than 100 BSc, MSc and PhD theses. Prof. Ancheyta has also been supervisor of a number of postdoctoral and sabbatical year professors.

Prof. Ancheyta has been working in the development and application of petroleum refining catalysts, kinetic and reactor models, and process technologies, mainly in catalytic cracking, catalytic reforming, middle distillate hydrotreating and heavy oils upgrading. He is author and co-author of a number of patents and books and about 200 scientific papers; he has been awarded the highest distinction (Level III) as National Researcher by the Mexican government and is a member of the Mexican Academy of Science. He has also been guest editor of various international journals, for example *Catalysis Today, Petroleum Science and Technology, Industrial Engineering Chemistry Research, Chemical Engineering Communications* and *Fuel*. Prof. Ancheyta has also chaired numerous international conferences.

Preface

Reaction kinetics is mainly focused on studying the rate at which chemical reactions proceed. It is also used to analyse the factors that affect the reaction rates and the mechanisms by means of which they take place.

The study of the chemical kinetics of a reaction is a fundamental tool to perform in the design of chemical reactors, to predict the reactor's performance and to develop new processes. In fact, the first step for designing a chemical reactor is always the generation of experimental data whereby the reaction rate expressions are determined.

Chemical Reaction Kinetics: Concepts, Methods and Case Studies is devoted to describing the fundamentals of reaction kinetics, with particular emphasis on the mathematical treatment of the experimental data. The book is organized in six chapters, each one having detailed deductions of the kinetic models with examples.

Chapter 1 deals with the definitions of the main concepts of stoichiometry, reacting systems, chemical kinetics and ideal reactors.

Chapter 2 gives details about the mathematical methods to determine the reaction order and the reaction rate coefficient for irreversible reactions with one component. The methods described here include the integral method, differential method, total pressure method and half-life time method.

Chapter 3 reports the mathematical methods for evaluating the kinetics of irreversible reactions with two or three components by employing the integral method, differential method and initial reaction rate method. All of the mathematical treatments are performed according to the type of feed composition: stoichiometric, non-stoichiometric and with a reactant in excess.

Chapter 4 describes the reversible reactions of first order, second order and combined orders.

Chapter 5 presents the mathematical treatment of complex reactions, that is, simultaneous or parallel irreversible reactions and consecutive or in-series irreversible reactions, with the same order or with combined orders.

Chapter 6 is devoted to special topics in kinetic modelling, which include reconciliation of data generated during experiments to minimize the inconsistencies of mass balances due to experimental errors, a method for sensitivity analysis to assure that kinetic parameters are properly estimated and the convergence of the objective function to the global minimum is achieved, estimation of kinetic parameters of enzymatic reactions by means of different approaches, estimation of kinetic parameters of catalytic cracking reaction using a lumping approach and estimation of kinetic parameters of hydrodesulphurization of petroleum distillates.

Each chapter illustrates the application of the different methods with detailed examples by using experimental information reported in the literature. Step-by-step solutions are provided so that the methods can be easily followed and applied for other situations. Some exercises are provided at the end to allow the reader to apply all of the methods developed in the previous chapters.

Chemical Reaction Kinetics: Concepts, Methods and Case Studies is oriented to cover the contents of undergraduate and postgraduate courses on reaction kinetics of chemical engineering and similar careers. It is anticipated that *Chemical Reaction Kinetics: Concepts, Methods and Case Studies* will become an outstanding and distinctive textbook because it emphasizes detailed description of fundamentals, mathematical treatments and examples of chemical reaction kinetics, which are not described with such details in previous textbooks related to the topic. The particular manner in which the kinetic models are developed will help the readers adapt to their own reaction studied and experimental data.

I would like to acknowledge Prof. Miguel A. Valenzuela from the School of Chemical Engineering and Extractive Industries at the National Polytechnic Institute of Mexico, who contributed some ideas during the preparation of the Spanish version of this book, and also to hundreds of students who during more than 20 years of delivering lectures encouraged me to write this book.

Jorge Ancheyta

1

Fundamentals of Chemical Reaction Kinetics

In homogeneous reacting systems, all the reactants and products are in the same phase. If the reaction involves a catalyst, it is also in the same phase. To determine the rates of reaction, experimental information is needed, which is generated by using properly designed small-scale reactors and experiments. These reaction rates cannot be directly measured, but they are obtained by means of experimental data such as the variation of time with respect to concentration of reactants or products, partial pressures and total pressure, among others.

To obtain the kinetic expression that represents the studied reaction, there are various approaches that correlate the experimental data with the variables that affect them.

When a reaction proceeds, one or more reactants can take part. It can be carried out in either liquid or gas phase, the reaction extent is measured by means of variations of reactants or product properties, or simply the reaction mechanisms are unknown. In any case, it is necessary before starting with the mathematical treatment of the experimental data to know the fundamentals of stoichiometry, thermodynamics and kinetics that will be further used for elucidating the specific mathematical expression for each type of reaction. This chapter is then devoted to introducing the readers to these topics.

1.1 Concepts of Stoichiometry

1.1.1 Stoichiometric Number and Coefficient

A chemical reaction can be represented as follows:

$$aA + bB + \ldots \rightarrow rR + sS + \ldots \tag{1.1}$$

where A, B, R and S are the chemical species, and a, b, r and s are their corresponding stoichiometric coefficients, which are the positive numbers before the chemical formula that balance the reaction.

Chemical Reaction Kinetics: Concepts, Methods and Case Studies, First Edition.
Jorge Ancheyta.
© 2017 John Wiley & Sons Ltd. Published 2017 by John Wiley & Sons Ltd.

Eq. (1.1) can be transformed as follows (Chopey, 1994):

$$-aA - bB - \ldots + rR + sS + \ldots = 0 \qquad (1.2)$$

or with positive values:

$$v_1A_1 + v_2A_2 + v_3A_3 + \ldots + v_{n-1}A_{n-1} + v_nA_n = 0 \qquad (1.3)$$

which can be generalized as:

$$\sum_{i=1}^{n} v_iA_i = 0 \qquad (1.4)$$

where A_i is the chemical formula and v_i is the corresponding stoichiometric numbers.

Stoichiometric numbers (v_i) are numerically equal to stoichiometric coefficients (a, b, r and s), but they have a negative sign for reactants and positive sign for products.

Example 1.1   Determine the stoichiometric coefficients and numbers for the following reaction for synthesis of ammonia:

$$N_2 + 3H_2 \rightarrow 2NH_3 \quad (aA + bB \rightarrow rR)$$

Solution

According to stoichiometry, the stoichiometric coefficients and numbers are:

Stoichiometric coefficient	Stoichiometric number
$a = 1$	$v_{N2} = -1$
$b = 3$	$v_{H2} = -3$
$r = 2$	$v_{NH3} = 2$

1.1.2   Molecularity

Molecularity is defined as the number of molecules of reactants that take part in a chemical reaction. Most of the reactions exhibit a molecularity of one or two, and in rare cases it reaches the value of three (Hill, 1977).

Molecularity is an appropriate concept for a process in which a simple or elemental step is occurring. Reactions in which one or several reactants produce one or several products in a simple path are scarce. For complex reactions, it is necessary to know the molecularity of each individual step of the reaction.

Table 1.1 Chemical reactions with different molecularity.

Molecularity	Examples	
1	$A \rightarrow R$	$n-C_4H_{10} \rightarrow i-C_4H_{10}$
	$A \rightarrow R + S$	$SO_2Cl_2 \rightarrow SO_2 + Cl_2$
	$A \rightarrow Products$	$(CH_3)_2O \rightarrow CH_4 + H_2 + CO$
2	$A + B \rightarrow R + S$	$O_3 + NO \rightarrow O_2 + NO_2$
	$A + B \rightarrow 2R$	$H_2 + Br_2 \rightarrow 2HBr$
	$A + B \rightarrow R$	$C_2H_4 + HI \rightarrow C_2H_5I$
	$2A \rightarrow Products$	$2N_2O \rightarrow 2N_2 + O_2$
3	$A + B + C \rightarrow Products$	$C_2H_5NO_2 + C_5H_5N + I_2 \rightarrow$ $C_2H_4INO_2 + C_5H_5NH^+ + I^-$
	$2A + B \rightarrow Products$	$2NO + O_2 \rightarrow 2NO_2$

Based on this concept, the chemical reactions can be classified mainly as mono-molecular, bi-molecular and tri-molecular. A mono-molecular reaction involves one molecule of reactant. In a bi-molecular reaction, two molecules of reactants (either the same or different) are combined to form one or more products. Tri-molecular reactions are rare since they need the simultaneous collision of three molecules to produce one or several products. Examples of the different types or reactions according to their molecularity are shown in Table 1.1.

1.1.3 Reaction Extent

To follow the performance of a chemical reaction, it is necessary to define a parameter which properly represents the conversion of the reactants. In 1920, De Donder (1920) introduced the concept of reaction extent (ξ), by considering that the change in the number of moles of the chemical species is directly related to the stoichiometric number as follows:

$$\frac{\Delta n_2}{\Delta n_1} = \frac{v_2}{v_1} \quad or \quad \frac{\Delta n_3}{\Delta n_1} = \frac{v_3}{v_1}$$

or, in differential form:

$$\frac{dn_2}{dn_1} = \frac{v_2}{v_1} \quad or \quad \frac{dn_2}{v_2} = \frac{dn_1}{v_1}$$

$$\frac{dn_3}{dn_1} = \frac{v_3}{v_1} \quad or \quad \frac{dn_3}{v_3} = \frac{dn_1}{v_1}$$

For all chemical species, these equations can be generalized in the following manner:

$$\frac{dn_1}{v_1} = \frac{dn_2}{v_2} = \frac{dn_3}{v_3} = \ldots = \frac{dn_i}{v_i} \tag{1.5}$$

Defining the parameter ξ, as the reaction extent:

$$\frac{dn_i}{v_i} = d\xi \tag{1.6}$$

The integration of Eq. (1.6) gives:

$$\int_{n_{io}}^{n_i} dn_i = v_i \int_0^{\xi} d\xi$$

$$\xi_i = \frac{n_i - n_{io}}{v_i} \tag{1.7}$$

It is then observed that if a moles of A_1 react with b moles of A_2 to produce r moles of A_{n-1} and s moles of A_n, the reaction extent ξ is equal to 1. Therefore, in general, it can be stated that ξ_a moles of A_1 react with ξ_b moles of A_2 to produce ξ_r moles of A_{n-1} and ξ_s moles of A_n.

1.1.4 Molar Conversion

The molar fractional conversion (x_i) is an intensive normalized parameter referred preferably to the limiting reactant; it is defined as the fraction of such a reactant that is transformed into products (Froment *et al.*, 2010):

$$x_i = \frac{\text{Moles of reactant "i" transformed}}{\text{Initial moles of reactant "i"}} = \frac{n_{io} - n_i}{n_{io}} \tag{1.8}$$

where $0 \leq x_i \leq 1$.

Subindex "$_o$" refers to the number of moles at zero time (i.e. the beginning of the reaction). Conversion can be correlated with reaction extent by means of Eqs. (1.7) and (1.8):

$$n_i = n_{io} + v_i \xi_i \tag{1.9}$$

$$n_i = n_{io} - n_{io} x_i \tag{1.10}$$

where:

$$\xi_i = -\frac{n_{io}}{v_i} x_i \tag{1.11}$$

The maximum reaction extent $(\xi_i{}^{max})$ can be calculated from Eq. (1.11) for the maximum conversion value $(x_i{}^{max} = 1)$:

$$\xi_i^{max} = -\frac{n_{io}}{v_i} \tag{1.12}$$

which implies that the minimum and maximum values of ξ_i are in the following range:

$$0 \leq \xi_i \leq -\frac{n_{io}}{v_i}$$

1.1.5 Types of Feed Composition in a Chemical Reaction

When a chemical reaction involves more than one reactant, the feed composition is different depending on the relative initial concentrations of the chemical species:

- *Stoichiometric feed composition*: This occurs when the ratio between the stoichiometric coefficients of the reactants is equal to the ratio between the amount of moles or the molar initial concentrations of reactants.
- *Non-stoichiometric feed composition*: This is when the ratio between the stoichiometric coefficients of the reactants is different from the ratio between the amount of moles or the molar initial concentrations of reactants.
- *Equimolar feed composition*: This is when the same amount of reactants are used at the beginning of the reaction to keep the ratio between the amount of moles or the molar initial concentration equal to unity no matter the stoichiometric coefficients of the reaction.
- *Reactant in excess*: This is when the ratio between the amount of moles or the molar initial concentrations of the reactants with respect to the limiting reactant is much higher than the ratio between the stoichiometric coefficients.

Some feed compositions can be considered close to the stoichiometric feed composition, and this happens when the ratio between the amount of moles or the molar initial concentrations of reactants is more or less the same as the ratio between the stoichiometric coefficients.

If at the beginning of the reaction there are inert components, although they are not reacting, they must be considered to define the type of feed composition.

Example 1.2 Define the different feed compositions for the following reaction of formation of nitrogen dioxide:

$$2NO + O_2 \rightarrow 2NO_2 \quad (2A + B \rightarrow 2R)$$

Solution

If a feed consists of 4 moles of NO and 2 moles of O_2, the ratio of moles between them is $n_{O2}/n_{NO} = 2/4 = 0.5$, and the ratio between stoichiometric coefficients is $b/a = 1/2 = 0.5$. Since $n_{O2}/n_{NO} = b/a$, the feed composition is stoichiometric.

If the feed consists of 3 moles of NO and 2 moles of O_2, the ratio of moles between them is $n_{O2}/n_{NO} = 2/3 = 0.66$, and the ratio between stoichiometric coefficients is $b/a = 1/2 = 0.5$. Since $n_{O2}/n_{NO} \neq b/a$, the feed composition is non-stoichiometric.

If the feed consists of 4 moles of NO and 4 moles of O_2, the ratio of moles between them is $n_{O2}/n_{NO} = 4/4 = 1$, and therefore the feed composition is equimolar. This feed composition is also non-stoichiometric since $n_{O2}/n_{NO} \neq b/a$.

If the feed consists of 1 mol of NO and 20 moles of O_2, the ratio of moles between them is $n_{O2}/n_{NO} = 20/1 = 20$, and the ratio between stoichiometric coefficients is $b/a = 1/2 = 0.5$. Since $n_{O2}/n_{NO} \gg b/a$, it is considered that reactant B (O_2) is in excess.

If the feed consists of 4 moles of NO and 1.8 moles of O_2, the ratio of moles between them is $n_{O2}/n_{NO} = 1.8/4 = 0.45$, and the ratio between stoichiometric coefficients is $b/a = 1/2 = 0.5$. Since $n_{O2}/n_{NO} \approx b/a$, the feed composition is assumed to be close to stoichiometric.

An equimolar feed composition would also be 4 moles of NO, 4 moles of O_2 and 4 inert moles.

1.1.6 Limiting Reactant

The limiting reactant is the chemical species that in a chemical reaction is consumed before all of the other reactants (Himmelblau, 1970). If the reaction is carried out with only one reactant, the limiting reactant concept does not have meaning since it is obvious that it is the limiting one.

For reactions between two or more components with stoichiometric feed composition, any of the reactants can be the limiting one since they are consumed at the same rate. For other feed compositions, the definition of *limiting reactant* will depend on such a composition and on the reaction stoichiometry.

To know the limiting reactant in certain reactions, the concept of *reaction extent* can be used according to the following definition: "the limiting reactant is the chemical species that has the lowest value of maximum reaction extent (ξ_i^{max})."

Example 1.3 Determine the limiting reactant if, in the following reaction, 5 moles of ethylene bromide (A) and 2 moles of potassium iodide (B) are fed:

$$C_2H_4Br_2 + 3KI \rightarrow C_2H_4 + 2KBr + KI_3 \quad (A + 3B \rightarrow R + 2S + T)$$

Solution

The number of moles of each reactant in the feed and the corresponding stoichiometric numbers are:

$$n_{Ao} = 5 \text{ moles}, n_{Bo} = 2 \text{ moles}, v_A = -1 \text{ and } v_B = -3.$$

According to Eq. (1.12):

$$\xi_A^{max} = -\frac{n_{Ao}}{v_A} = -\frac{5}{-1} = 5$$

$$\xi_B^{max} = -\frac{n_{Bo}}{v_B} = -\frac{2}{-3} = {}^2/_3$$

Since $\xi_B^{max} < \xi_A^{max}$, thus the limiting reactant is B.

The results of this example can be confirmed by analysing the stoichiometry of the reaction, in which for each mole of A, 3 moles of B are required. For the case of this example, if 5 moles of A are used, then 15 moles of B will be required to complete the reaction, and if only 2 moles of B are present, then B is consumed first and A is in excess.

1.1.7 Molar Balance in a Chemical Reaction

If, in the reaction $aA + bB \rightarrow rR + sS$, A is assumed to be the limiting reactant, and n_{Ao}, n_{Bo}, n_{Ro} and n_{So} are the number of moles of A, B, R and S, respectively, at the beginning of the reaction, then from Eq. (1.7) for reactant A:

$$n_A = n_{Ao} + v_A \xi_A$$

Substituting Eq. (1.11) in this equation:

$$n_A = n_{Ao} + v_A \left(-\frac{n_{Ao} x_A}{v_A} \right) = n_{Ao} - n_{Ao} x_A = n_{Ao}(1 - x_A) \qquad (1.13)$$

For reactant B:

$$n_B = n_{Bo} + v_B \xi_A$$

$$n_B = n_{Bo} + v_B \left(-\frac{n_{Ao} x_A}{v_A} \right)$$

Since $v_B = -b$ and $v_A = -a$, then:

$$n_B = n_{Bo} + (-b) \left(-\frac{n_{Ao} x_A}{-a} \right) = n_{Bo} - \frac{b}{a} n_{Ao} x_A \qquad (1.14)$$

which can also be written as follows to introduce the ratio n_{Bo}/n_{Ao}:

$$n_B = n_{Ao} \left(\frac{n_{Bo}}{n_{Ao}} - \frac{b}{a} x_A \right)$$

Defining the following feed molar ratio of B with respect to A:

$$M_{BA} = {}^{n_{Bo}}\big/_{n_{Ao}}$$

$$n_B = n_{Ao}\left(M_{BA} - \frac{b}{a}x_A\right) \tag{1.15}$$

Following the same procedure for R and S:

$$n_R = n_{Ro} + \frac{r}{a}n_{Ao}x_A \tag{1.16}$$

$$n_R = n_{Ao}\left(M_{RA} + \frac{r}{a}x_A\right) \tag{1.17}$$

$$n_S = n_{So} + \frac{s}{a}n_{Ao}x_A \tag{1.18}$$

$$n_S = n_{Ao}\left(M_{SA} + \frac{s}{a}x_A\right) \tag{1.19}$$

where:

$$M_{RA} = {}^{n_{Ro}}\big/_{n_{Ao}} \text{ and } M_{SA} = {}^{n_{So}}\big/_{n_{Ao}}$$

1.1.8 Relationship between Conversion and Physical Properties of the Reacting System

When it is not possible to generate experimental information in terms of common properties (concentration, total pressure, partial pressure, etc.), it is necessary to measure the reaction extent as a function of any physical property of the system, such as absorbance, electric conductivity, refractive index and the like, since they are additive functions of the contributions of all chemical species and in general they vary linearly with the concentration (Levenspiel, 1972).

For any physical property (λ), the contribution of all the chemical species can be represented by:

$$\lambda = \sum_{i=1}^{n} y_i \lambda_i \tag{1.20}$$

The relationship between λ and C_i can be written as:

$$\lambda_i \propto C_i \quad \text{or} \quad \lambda_i = k_{\lambda i} C_i \tag{1.21}$$

Dividing Eq. (1.7) between the volume to obtain the volumetric reaction extent (ξ_i'):

$$\xi_i' = \frac{\xi_i}{V} = \frac{n_i - n_{io}}{V v_i} = \frac{\dfrac{n_i}{V} - \dfrac{n_{io}}{V}}{v_i} = \frac{C_i - C_{io}}{v_i}$$

$$C_i = C_{io} + v_i \xi_i'$$

And, substituting in Eq. (1.21):

$$\lambda_i = k_{\lambda i}\left(C_{io} + v_i \xi_i'\right) = k_{\lambda i} C_{io} + k_{\lambda i} v_i \xi_i'$$

Substituting λ_i in Eq. (1.20) and considering a constant value of ξ_i' for a specific chemical species:

$$\lambda = \sum_{i=1}^{n} y_i \left(k_{\lambda i} C_{io} + k_{\lambda i} v_i \xi_i'\right) = \sum_{i=1}^{n} y_i k_{\lambda i} C_{io} + \xi_i' \sum_{i=1}^{n} y_i k_{\lambda i} v_i \qquad (1.22)$$

Since $k_{\lambda i}$ and v_i are constant, the following equation can be derived:

$$\sum_{i=1}^{n} y_i k_{\lambda i} v_i = k_{\lambda i} v_i \sum_{i=1}^{n} y_i = k_{\lambda i} v_i = K_\lambda$$

Moreover, at zero time, Eq. (1.20) is:

$$\lambda_o = \sum_{i=1}^{n} y_i \lambda_{io} = \sum_{i=1}^{n} y_i k_{\lambda i} C_{io}$$

Substituting K_λ and λ_o in Eq. (1.22):

$$\lambda = \lambda_o + K_\lambda \xi_i' \quad \text{or} \quad \lambda - \lambda_o = K_\lambda \xi_i' \qquad (1.23)$$

Applying Eq. (1.23) at the maximum point of reaction extent:

$$\lambda_\infty - \lambda_o = K_\lambda \xi_i'^{max} \qquad (1.24)$$

Dividing Eq. (1.23) by Eq. (1.24):

$$\frac{\lambda - \lambda_o}{\lambda_\infty - \lambda_o} = \frac{K_\lambda \xi_i'}{K_\lambda \xi_i'^{max}} = \frac{\xi_i'}{\xi_i'^{max}} = \frac{\xi_i/V}{\xi_i^{max}/V} = \frac{\xi_i}{\xi_i^{max}} \qquad (1.25)$$

Since the maximum reaction extent (ξ_i^{max}) is:

$$\xi_i^{max} = -\frac{n_{io}}{v_i}$$

the ratio (ξ_i/ξ_i^{max}) is:

$$\frac{\xi_i}{\xi_i^{max}} = \frac{\left(-n_{io}/v_i\right) x_i}{\left(-n_{io}/v_i\right)} = x_i \qquad (1.26)$$

And, finally, Eq. (1.25) is:

$$x_i = \frac{\lambda - \lambda_o}{\lambda_\infty - \lambda_o} \qquad (1.27)$$

where:

λ: Physical property at time t;
λ_o: Physical property at time zero ($t = 0$);
λ_∞: Physical property that does not change with time; and
x_i: Conversion.

Example 1.4   Reactant A is prepared under refrigeration and is introduced in a small capillary that acts as a reaction vessel, in which the decomposition reaction $A \rightarrow R + S$ is carried out. The vessel is rapidly introduced in a bath containing water at the boiling point. During handling, there is no reaction. During the experiments, several data of the capillary length occupied by the reacting mixture (L) were collected (Levenspiel, 1979). Evaluate the values of conversion for the capillary length at different times indicated in Table 1.2.

Solution

In this case, Eq. (1.27) can be written as:

$$x_A = \frac{L - L_o}{L_\infty - L_o}$$

where:

L: Capillary length at time t;
L_o: Capillary length at time zero ($t = 0$);
L_∞: Capillary length that does not change with time; and
x_A: Conversion of reactant A.

In this equation, the initial capillary length (L_o) is unknown. However, from the analysis of the stoichiometry, it is deduced that the reaction is

Table 1.2 Data and results of Example 1.4.

Time (min)	Capilar length (cm)	x_A
0.5	6.1	0.2979
1.0	6.8	0.4468
1.5	7.2	0.5319
2.0	7.5	0.5957
3.0	7.85	0.6702
4.0	8.1	0.7234
6.0	8.4	0.7872
10.0	8.7	0.8511
∞	9.4	1.0000

irreversible; hence, at $t = \infty$, the conversion is 100% ($x_A = 1.0$), that is, all the reactant A has been transformed into R and S. In other words, 1 mole of reactant has been transformed into 2 moles of products. This indicates that at $t = \infty$, the number of moles is duplicated, as well as the volume and the capillary length, so that:

$$L_\infty = 2 L_o$$

$$L_o = L_\infty / 2 = 9.4 / 2 = 4.7 cm$$

The application of the previous equation, at $t = 1$ min, gives:

$$x_A = \frac{L - L_o}{L_\infty - L_o} = \frac{6.8 - 4.7}{9.4 - 4.7} = 0.2979$$

The results of conversion for all the capillary lengths are reported in Table 1.2.

1.2 Reacting Systems

1.2.1 Mole Fraction, Weight Fraction and Molar Concentration

If the total number of moles and weight of all the chemical species present in the reacting mixture are n_t and w_t, respectively, and if n_i moles and w_i weight units of component i are present, the mole or molar fraction (y_i) and the weight fraction (y_{wi}) of species i in the system are defined as:

$$y_i = \frac{n_i}{n_t} \quad \text{(mole fraction)} \tag{1.28}$$

$$y_{wi} = \frac{w_i}{w_t} \quad \text{(weight fraction)} \tag{1.29}$$

By definition, the sum of fractions of all the components must be equal to unity:

$$\sum_{i=1}^{n} y_i = y_1 + y_2 + \ldots + y_n = \frac{n_1}{n_t} + \frac{n_2}{n_t} + \ldots + \frac{n_n}{n_t} = \frac{n_1 + n_2 + \ldots + n_n}{n_t} = \frac{n_t}{n_t} = 1$$

$$\sum_{i=1}^{n} y_{wi} = y_{w1} + y_{w2} + \ldots + y_{wn} = \frac{w_1}{w_t} + \frac{w_2}{w_t} + \ldots + \frac{w_n}{w_t} = \frac{w_1 + w_2 + \ldots + w_n}{w_t} = \frac{w_t}{w_t} = 1$$

To convert a mole fraction in a weight fraction or vice versa, the following relationship is used, which is obtained by using the definition of number of moles ($n = w/MW$):

$$y_i = \frac{n_i}{n_t} = \frac{w_i / MW_i}{w_t / MW_t} = \frac{w_i MW_t}{w_t MW_i} = y_{wi} \left(\frac{MW_t}{MW_i} \right) \tag{1.30}$$

where the molecular weight of the mixture (MW_t) is:

$$MW_t = \sum_{i=1}^{n} y_i MW_i \qquad (1.31)$$

The molar concentration is defined as the ratio between the number of moles of a chemical species (n_i) by unit of the system volume (V) and is related with density (ρ_i) as follows:

$$C_i = \frac{n_i}{V} = \frac{w_i/PM_i}{V} = \frac{w_i}{V\,PM_i} = \frac{\rho_i}{PM_i} \qquad (1.32)$$

Example 1.5 Evaluate the initial mole and weight fractions of the reactants if the following reaction of nitrogen dioxide formation starts with 3 moles of NO and 2 moles of O_2.

$$2NO + O_2 \rightarrow 2NO_2 \quad (2A + B \rightarrow 2R)$$

Solution

Mole fractions. Using Eq. (1.28):

$$y_{NOo} = \frac{n_{NOo}}{n_{to}} = \frac{n_{NOo}}{n_{NOo} + n_{O2o}} = \frac{3}{3+2} = 0.6$$

$$y_{O2o} = \frac{n_{O2o}}{n_{to}} = \frac{n_{O2o}}{n_{NOo} + n_{O2o}} = \frac{2}{3+2} = 0.4$$

Weight fractions. Using Eq. (1.29):

$$w_{NOo} = (n_{NOo})(MW_{NO}) = (3)(30) = 90\ g$$

$$w_{O2o} = (n_{O2o})(MW_{O2}) = (2)(32) = 64\ g$$

$$w_{to} = w_{NOo} + w_{O2o} = 90 + 64 = 154\ g$$

$$y_{wNOo} = \frac{w_{NOo}}{w_{to}} = \frac{90}{154} = 0.584$$

$$y_{wO2o} = \frac{w_{O2o}}{w_{to}} = \frac{64}{154} = 0.416$$

Using Eq. (1.30):

$$MW_t = y_{NO}MW_{NO} + y_{O2}MW_{O2} = 0.6(30) + 0.4(32) = 30.8\ g/gmol$$

$$y_{wNOo} = y_{NOo}\left(\frac{MW_{NO}}{MW_t}\right) = 0.6\left(\frac{30}{30.8}\right) = 0.584$$

$$y_{wO2o} = y_{O2o}\left(\frac{MW_{O2}}{MW_t}\right) = 0.4\left(\frac{32}{30.8}\right) = 0.416$$

1.2.2   Partial Pressure

Partial pressure is defined as the pressure that a gas in a mixture of gases would exert if it alone occupied the whole volume occupied by the mixture at the same temperature. Therefore, partial pressure p_i of gas i in a mixture of gases is calculated by multiplying its mole fraction (y_i) by the total pressure of the system (P):

$$p_i = y_i P$$

This is the so-called Dalton law, which also states that the total pressure exerted of a mixture of gases is equal to the sum of partial pressures of all the gases of the mixture (Smith *et al.*, 1980):

$$\sum_{i=1}^{n} p_i = \sum_{i=1}^{n} y_i P = P \sum_{i=1}^{n} y_i = P$$

Partial pressure of a gas in a mixture of gases is related to its molar concentration by means of the ideal gas law:

$$p_i V = n_i RT$$
$$p_i = y_i P = \left(\frac{n_i}{V}\right) RT = C_i RT \tag{1.33}$$

where R is the universal gas constant, the common values of which are:

$$R = 1.987 \frac{Cal}{gmol\ K} = 1.986 \frac{BTU}{lbmol\ R} = 82.057 \frac{atm\ cm^3}{gmol\ K} = 8.314 \frac{J}{gmol\ K} = 0.08205 \frac{atm\ lt}{gmol\ K}$$

$$= 10.73 \frac{psia\ ft^3}{gmol\ K} 0.7302 \frac{atm\ ft^3}{lbmol\ R} = 62.361 \frac{mmHg\ lt}{gmol\ K} = 1.315 \frac{atm\ ft^3}{lbmol\ K} = 8.31 \frac{KPa\ lt}{gmol\ K}$$

$$5.83 \times 10^{-4} \frac{KWh}{lbmol\ R} = 7.82 \times 10^{-4} \frac{hp\ h}{lbmol\ R}$$

1.2.3   Isothermal Systems at Constant Density

When a system operates at constant density, the corresponding volume refers to the reacting mixture and not to the volume of the reactor. To this type of system belongs those reactions conducted in liquid phase or gas phase that either do not experience change in the number of moles or are carried out in hermetic vessels.

For an homogeneous reaction, which is carried out in the hermetic vessel shown in Figure 1.1, in gas phase, isothermally, with change in the number of moles and consequently with an increase or decrease in the

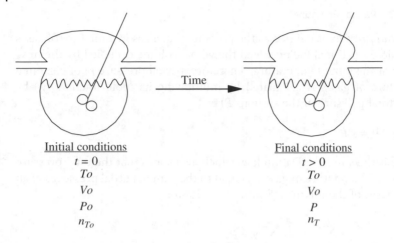

Figure 1.1 Reacting system at constant density.

pressure by expansion or compression due to the reaction, the molar balance at any time for the reaction $aA + bB \rightarrow rR + sS$ is the following:

$$n_A = n_{Ao} - n_{Ao}x_A$$

$$n_B = n_{Bo} - \frac{b}{a}n_{Ao}x_A$$

$$n_R = n_{Ro} + \frac{r}{a}n_{Ao}x_A$$

$$n_S = n_{So} + \frac{s}{a}n_{Ao}x_A$$

Assuming that the system follows the ideal gas law ($PV = nRT$) for the system at constant density, then at the beginning of the reaction (Figure 1.1):

$$P_o V_o = n_{To}RT_o \tag{1.34}$$

$$n_{To} = n_{Ao} + n_{Bo} + n_{Ro} + n_{So} \tag{1.35}$$

and at time t:

$$PV_o = n_T RT_o \tag{1.36}$$

$$n_T = n_A + n_B + n_R + n_S \tag{1.37}$$

Substituting the equations of the molar balance (Eqs. 1.13, 1.15, 1.17 and 1.19) in Eq. (1.37):

$$n_T = n_{Ao} + n_{Bo} + n_{Ro} + n_{So} + \frac{r}{a}n_{Ao}x_A + \frac{s}{a}n_{Ao}x_A - n_{Ao}x_A - \frac{b}{a}n_{Ao}x_A$$

$$n_T = n_{To} + \frac{n_{Ao}}{a}(r+s-a-b)x_A = n_{To} + \frac{n_{Ao}\Delta n}{a}x_A$$

$$\tag{1.38}$$

where:

$$\Delta n = r + s - a - b$$

Dividing Eq. (1.36) by Eq. (1.34):

$$\frac{PV_o}{P_o V_o} = \frac{n_T R T_o}{n_{To} R T_o}$$

$$\frac{P}{P_o} = \frac{n_T}{n_{To}} \tag{1.39}$$

Substituting Eq. (1.38) in (1.39):

$$\frac{P}{P_o} = \frac{n_{To} + \dfrac{n_{Ao}\Delta n}{a} x_A}{n_{To}} = 1 + \left(\frac{n_{Ao}}{n_{To}}\right)\frac{\Delta n}{a} x_A = 1 + \left(\frac{y_{Ao}\Delta n}{a}\right) x_A = 1 + \varepsilon_A x_A$$

where ε_A is the molar expansion factor:

$$\varepsilon_A = \frac{y_{Ao}\Delta n}{a} \tag{1.40}$$

ε_A is calculated with two parameters: stoichiometry (Δn, a) and feed composition (y_{Ao}). ε_A becomes important if the density of the reacting system is changing as the reaction proceeds.

The common forms of the previous equations are:

$$P = P_o(1 + \varepsilon_A x_A) \tag{1.41}$$

$$x_A = \frac{P - P_o}{\varepsilon_A P_o} \tag{1.42}$$

Concentrations at any time in a system at constant density are calculated by dividing the equations of the molar balance by the volume of the reacting system, which results in:

$$C_A = C_{Ao} - C_{Ao} x_A = C_{Ao}(1 - x_A) \tag{1.43}$$

$$C_B = C_{Bo} - \frac{b}{a} C_{Ao} x_A = C_{Ao}\left(M_{BA} - \frac{b}{a} x_A\right) \tag{1.44}$$

$$C_R = C_{Ro} + \frac{r}{a} C_{Ao} x_A = C_{Ao}\left(M_{RA} + \frac{r}{a} x_A\right) \tag{1.45}$$

$$C_S = C_{So} + \frac{s}{a} C_{Ao} x_A = C_{Ao}\left(M_{SA} + \frac{s}{a} x_A\right) \tag{1.46}$$

where M_{iA} are the initial molar ratios of B, R and S with respect to the limiting reactant A, and are given by:

$$M_{BA} = {C_{Bo}}\big/{C_{Ao}} = {n_{Bo}}\big/{n_{Ao}}$$

$$M_{RA} = {C_{Ro}}\big/{C_{Ao}} = {n_{Ro}}\big/{n_{Ao}}$$

$$M_{SA} = {C_{So}}\big/{C_{Ao}} = {n_{So}}\big/{n_{Ao}}$$

1.2.3.1 Relationship between Partial Pressure (p_A) and Conversion (x_A)

Substitution of Eq. (1.43) in Eq. (1.33) gives:

$$p_A = \left(\frac{n_A}{V}\right)RT = C_A RT = C_{Ao}(1 - x_A)RT \tag{1.47}$$

At the beginning of the reaction ($t = 0$, $x_A = 0$, $C_A = C_{Ao}$), Eq. (1.47) is:

$$p_{Ao} = C_{Ao}RT \tag{1.48}$$

Dividing Eq. (1.47) by Eq. (1.48):

$$\frac{p_A}{p_{Ao}} = \frac{C_{Ao}(1 - x_A)RT}{C_{Ao}RT} = (1 - x_A) \tag{1.49}$$

$$p_A = p_{Ao}(1 - x_A)$$

$$x_A = \frac{p_{Ao} - p_A}{p_{Ao}} \tag{1.50}$$

1.2.3.2 Relationship between Partial Pressure (p_A) and Total Pressure (P)

Substitution of Eq. (1.42) in Eq. (1.50) gives:

$$\frac{P - P_o}{\varepsilon_A P_o} = \frac{p_{Ao} - p_A}{p_{Ao}}$$

$$p_A = p_{Ao}\left(1 - \frac{P - P_o}{\varepsilon_A P_o}\right) = \frac{p_{Ao}}{\varepsilon_A P_o}(\varepsilon_A P_o - P + P_o)$$

From the Dalton law, $y_{Ao} = p_{Ao}/P_o$, thus:

$$p_A = \frac{y_{Ao}}{\varepsilon_A}[P_o(1 + \varepsilon_A) - P] = \frac{a}{\Delta n}[P_o(1 + \varepsilon_A) - P] \tag{1.51}$$

1.2.3.3 Relationship between Molar Concentration (C_A) and Total Pressure (P)

Substitution of Eq. (1.42) in Eq. (1.43) gives:

$$C_A = C_{Ao}\left(1 - \frac{P - P_o}{\varepsilon_A P_o}\right) = \frac{C_{Ao}}{\varepsilon_A P_o}[\varepsilon_A P_o + P_o - P]$$

Since $C_{Ao} = \dfrac{p_{Ao}}{RT} = \dfrac{P_o y_{Ao}}{RT}$:

$$C_A = \frac{C_{Ao}}{\varepsilon_A P_o}[P_o(1 + \varepsilon_A) - P] = \frac{y_{Ao}}{RT\varepsilon_A}[P_o(1 + \varepsilon_A) - P] \tag{1.52}$$

Example 1.6 In the decomposition reaction of ethylene oxide in gas phase conducted at 687 K in a hermetic vessel, the experimental data reported in Table 1.3 were obtained (Fogler, 1992). Calculate the conversion, concentrations and partial pressures of all the chemical species.

$$C_2H_4O \rightarrow CH_4 + CO\,(A \rightarrow R + S)$$

Solution

At the beginning of the reaction ($t = 0$), the initial total pressure (P_o) is 116.5 mmHg. Since only ethylene oxide is present, $y_{Ao} = 1$, so that the initial total pressure is the same as that of the ethylene oxide partial pressure ($P_o = p_{Ao} = 116.5$ mmHg). Thus, the initial concentration of ethylene oxide is:

$$C_{Ao} = \frac{p_{Ao}}{RT} = \frac{P_o y_{Ao}}{RT} = \frac{(116.5\ mmHg)(1)}{\left(62.361\dfrac{mmHg\ lt}{gmolK}\right)(687K)} = 2.719 \times 10^{-3}\frac{gmol}{lt}$$

The molar expansion factor is:

$$\varepsilon_A = \frac{y_{Ao}\Delta n}{a} = \frac{(1)(2-1)}{1} = 1$$

The calculation for $P = 112.6$ mmHg and $t = 5$ min is as follows. Data of total pressure are changed to concentration with Eq. (1.52):

$$C_A = \frac{y_{Ao}}{RT\varepsilon_A}[P_o(1+\varepsilon_A)-P] = \frac{1}{\left(62.361\dfrac{mmHg\ lt}{gmolK}\right)(687K)(1)}[116.5(1+1)-P]$$

$$C_A = \frac{y_{Ao}}{RT\varepsilon_A}[P_o(1+\varepsilon_A)-P] = \frac{1}{(42842.007)}(233-122.6) = 2.577 \times 10^{-3}\frac{gmol}{lt}$$

Table 1.3 Data and results of Example 1.6.

t (min)	P (mmHg)	x_A	$C_A \times 10^3$ (gmol/lt)	$C_R = C_S \times 10^4$ (gmol/lt)	p_A (mmHg)	$p_R = p_S$ (mmHg)
0	116.5	0	2.719	0	116.5	0
5	122.6	0.0523	2.577	1.424	110.4	6.1
7	125.7	0.0789	2.505	2.147	107.3	9.2
9	128.7	0.1047	2.435	2.848	104.3	12.2
12	133.2	0.1433	2.329	3.898	99.8	16.7
18	141.2	0.2120	2.143	5.765	91.8	24.7

To calculate the partial pressure of A (p_A), Eqs. (1.33) or (1.51) can be used. With Eq. (1.33):

$$p_A = C_A RT = (2.577 \times 10^{-3})(62.361)(687) = 110.4 \; mmHg$$

With Eq. (1.51):

$$p_A = \frac{a}{\Delta n}[P_o(1 + \varepsilon_A) - P] = \frac{1}{2-1}[116.5(1+1) - 122.6] = 110.4 \; mmHg$$

Conversion can be determined with Eqs. (1.42), (1.43) or (1.50). With Eq. (1.42):

$$x_A = \frac{P - P_o}{\varepsilon_A P_o} = \frac{122.6 - 116.5}{(1)(116.5)} = 0.0523$$

With Eq. (1.43):

$$x_A = \frac{C_{Ao} - C_A}{C_{Ao}} = \frac{2.719 \times 10^{-3} - 2.577 \times 10^{-3}}{2.719 \times 10^{-3}} = 0.0523$$

With Eq. (1.50):

$$x_A = \frac{p_{Ao} - p_A}{p_{Ao}} = \frac{116.5 - 110.4}{116.5} = 0.0523$$

Concentrations of products are computed with Eqs. (1.45) and (1.46). Since R and S are not present at the beginning of the reaction, the initial molar ratios M_{RA} and M_{SA} are both equal to zero. In addition, according to stoichiometry $r/a = s/a = 1$, so that:

$$C_R = C_{Ao} x_A$$
$$C_S = C_{Ao} x_A$$
$$C_R = C_S = C_{Ao} x_A = (2.719 \times 10^{-3})(0.0523) = 1.424 \times 10^{-4} \, gmol/lt$$

Partial pressures of products are calculated with Eq. (1.33):

$$p_R = p_S = C_R RT = C_S RT = (1.424 \times 10^{-4})(62.361)(687) = 6.1 \, mmHg$$

The results for the other values of total pressures are summarized in Table 1.3.

1.2.4 Isothermal Systems at Variable Density

For the gas phase reaction $aA + bB \rightarrow rR + sS$ (conducted at constant temperature and pressure in the vessel shown in Figure 1.2), in which a change in the number of moles occurs and hence an increase or decrease in the volume of the reacting system due to reaction, the molar balance is given by Eqs. (1.13)–(1.19).

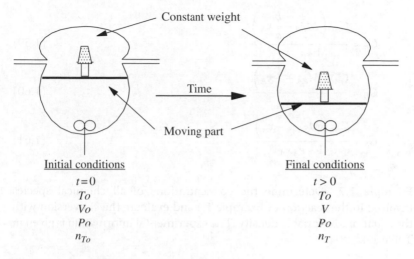

Figure 1.2 Reacting system at variable density.

At the beginning of the reaction:

$$P_o V_o = n_{To} R T_o \qquad (1.53)$$

At time t:

$$P_o V = n_T R T_o \qquad (1.54)$$

Dividing Eq. (1.54) by Eq. (1.53):

$$\frac{P_o V}{P_o V_o} = \frac{n_T R T_o}{n_{To} R T_o}$$

$$\frac{V}{V_o} = \frac{n_T}{n_{To}} \qquad (1.55)$$

Substitution of Eq. (1.38) in Eq. (1.55) gives:

$$\frac{V}{V_o} = \frac{n_{To} + \dfrac{n_{Ao} \Delta n}{a} x_A}{n_{To}} = 1 + \varepsilon_A x_A \qquad (1.56)$$

$$V = V_o (1 + \varepsilon_A x_A)$$

$$x_A = \frac{V - V_o}{\varepsilon_A V_o} \qquad (1.57)$$

To calculate the concentrations at any time, the equations of the molar balance (Eqs. 1.13, 1.15, 1.17 and 1.19) are divided by Eq. (1.56):

$$\frac{n_A}{V} = \frac{n_{Ao}}{V_o} \frac{(1 - x_A)}{(1 - \varepsilon_A x_A)}$$

$$C_A = \frac{C_{Ao}(1 - x_A)}{1 + \varepsilon_A x_A} \qquad (1.58)$$

$$C_B = \frac{C_{Ao}\left(M_{BA} - \dfrac{b}{a}x_A\right)}{1 + \varepsilon_A x_A} \tag{1.59}$$

$$C_R = \frac{C_{Ao}\left(M_{RA} + \dfrac{r}{a}x_A\right)}{1 + \varepsilon_A x_A} \tag{1.60}$$

$$C_S = \frac{C_{Ao}\left(M_{SA} + \dfrac{s}{a}x_A\right)}{1 + \varepsilon_A x_A} \tag{1.61}$$

Example 1.7 Determine the concentrations of all chemical species involved in the reaction of Example 1.4 and evaluate the conversion with the equations at variable density. The experimental information is given in Table 1.4.

Solution

The stoichiometry of the reaction is $A \rightarrow R + S$, and the reactant A is fed pure. Similarly to Example 1.4, L_o is unknown. At the beginning of the reaction and at any time:

$$At \quad t = 0, \; V_o = A L_o$$
$$At \quad t > 0, V = A L$$

From Eq. (1.56):

$$V = V_o(1 + \varepsilon_A x_A)$$

Table 1.4 Data and results of Example 1.7.

Time (min)	Capilar length (cm)	x_A	$C_A \times 10^{-2}$ (gmol/lt)	$C_R \times 10^{-2} = C_s \times 10^{-2}$ (gmol/lt)
0.5	6.1	0.2979	4.091	1.736
1.0	6.8	0.4468	2.892	2.336
1.5	7.2	0.5319	2.311	2.626
2.0	7.5	0.5957	1.916	2.824
3.0	7.85	0.6702	1.494	3.035
4.0	8.1	0.7234	1.214	3.175
6.0	8.4	0.7872	0.901	3.332
10.0	8.7	0.8511	0.608	3.478
∞	9.4	1.0000	0.000	3.782

so that:

$$AL = AL_o(1 + \varepsilon_A x_A)$$
$$L = L_o(1 + \varepsilon_A x_A)$$

The molar expansion factor is:

$$\varepsilon_A = \frac{y_{Ao}\Delta n}{a} = \frac{(1)(2-1)}{1} = 1$$

then:

$$L = L_o(1 + x_A)$$

At $t = \infty$:

$$L = L_\infty \text{ and } x_A = 1.0$$
$$L_\infty = L_o(1 + 1) = 2L_o$$

and, therefore:

$$L_o = \frac{L_\infty}{2} = \frac{9.4}{2} = 4.7$$

This result is the same as that calculated in Example 1.4; thus, the conversion values are calculated with Eq. (1.27) using the values of capillary length, which are summarized in Table 1.4.

During reaction, the total pressure is kept constant, and is equal to the atmospheric pressure plus the exerted pressure by mercury:

$$P = P_{atm} + 1000\,mmHg = 1760\,mmHg$$

Since the reaction vessel is immersed in a bath with boiling water, the temperature of the reaction is:

$$T = 100°C + 273.15 = 373.15\,K$$

The initial concentration of the reactant is then:

$$C_{Ao} = \frac{p_{Ao}}{RT} = \frac{P_o y_{Ao}}{RT} = \frac{(1760\,mmHg)(1)}{\left(62.361\dfrac{mmHg\,lt}{gmolK}\right)(373.15K)} = 7.563 \times 10^{-2}\frac{gmol}{lt}$$

Using Eqs. (1.58), (1.60) and (1.61), the values of C_A, C_R and C_S can be calculated.

Since A is fed pure, the initial molar ratios are zero, $M_{RA} = M_{SA} = 0$; moreover, according to the stoichiometry, the ratio between stoichiometric coefficients is $r/a = s/a = 1$, so that the concentrations of R and S are the same at any time, $C_R = C_S$. For instance, at $t = 0.5$ min and $x_A = 0.2979$:

$$C_A = \frac{C_{Ao}(1-x_A)}{1 + \varepsilon_A x_A} = \frac{7.563 \times 10^{-2}(1-0.2979)}{1 + (1)(0.2979)} = 4.091 \times 10^{-2} \frac{gmol}{lt}$$

$$C_R = C_s = \frac{C_{Ao}\left(M_{RA} + \dfrac{r}{a}x_A\right)}{1 + \varepsilon_A x_A} = \frac{C_{Ao}x_A}{1 + x_A} = \frac{(7.563 \times 10^{-2})(0.2979)}{1 + 0.2979}$$

$$= 1.736 \times 10^{-2} \frac{gmol}{lt}$$

The complete results are reported in Table 1.4.

1.2.5 General Case of Reacting Systems

The reacting systems working at constant temperature either at constant density or at variable density, as described in this chapter, are the most frequently used to generate kinetic data about the variation of certain properties with respect to time in batch reactors.

For the general case, in which temperature, pressure and density change, the equations to calculate the concentration of all the chemical species are:

$$C_A = \frac{C_{Ao}(1-x_A)}{1 + \varepsilon_A x_A}\left(\frac{T_o}{T}\right)\left(\frac{P}{P_o}\right) \tag{1.62}$$

$$C_B = \frac{C_{Ao}\left(M_{BA} - \dfrac{b}{a}x_A\right)}{1 + \varepsilon_A x_A}\left(\frac{T_o}{T}\right)\left(\frac{P}{P_o}\right) \tag{1.63}$$

$$C_R = \frac{C_{Ao}\left(M_{RA} + \dfrac{r}{a}x_A\right)}{1 + \varepsilon_A x_A}\left(\frac{T_o}{T}\right)\left(\frac{P}{P_o}\right) \tag{1.64}$$

$$C_S = \frac{C_{Ao}\left(M_{SA} + \dfrac{s}{a}x_A\right)}{1 + \varepsilon_A x_A}\left(\frac{T_o}{T}\right)\left(\frac{P}{P_o}\right) \tag{1.65}$$

In these equations, the subindex "$_o$" refers to the property at the beginning of the reaction, that is, at zero time ($t = 0$).

When the behaviour of the system cannot be represented by the ideal gas law, the previous equations must be multiplied by the ratio of compressibility factors (z_o/z).

1.2.6 Kinetic Point of View of the Chemical Equilibrium

A chemical reaction can finish due to kinetic restrictions or because it has reached the chemical equilibrium. By means of thermodynamics, it is possible to know the concentration of the different components at the equilibrium, and thus the maximum conversion or equilibrium conversion (Smith *et al.*, 1980).

For a reversible reaction, with forward and reverse elemental reactions $(aA + bB \Leftrightarrow rR + rS)$, the equilibrium is established when the reaction rates in both directions are equal:

$$(-r_A) = k_1 C_A^a C_B^b$$

$$(r_R) = k_2 C_R^r C_S^s$$

$$k_1 C_A^a C_B^b = k_2 C_R^r C_S^s$$

where k_1 is the forward reaction rate coefficient $(aA + bB \rightarrow rR + sS)$, and k_2 the reverse reaction rate coefficient $(rR + sS \rightarrow aA + bB)$. Introducing the equilibrium constant $K_e = k_1/k_2$:

$$K_e = K_c = \frac{k_1}{k_2} = \frac{C_R^r C_S^s}{C_A^a C_B^b} \tag{1.66}$$

According to thermodynamic laws, the equilibrium constant can be calculated with the Gibbs free energy of the species that take part during the reaction, which are values that are reported in the literature for most of the components.

If the reactants and products are gases, the ideal gas law can be used $(p_i = C_i RT)$ and Eq. (1.66) becomes:

$$K_c = \left(\frac{p_R^r p_S^s}{p_A^a p_B^b}\right)(RT)^{-\Delta n} = K_p(RT)^{-\Delta n} \tag{1.67}$$

where:

$$K_p = \left(\frac{p_R^r p_S^s}{p_A^a p_B^b}\right)$$

Combining Eq. (1.67) with Eq. (1.33):

$$K_p = \left(\frac{y_R^r y_S^s}{y_A^a y_B^b}\right)P^{\Delta n} = K_y P^{\Delta n} \tag{1.68}$$

where:

$$K_y = \left(\frac{y_R^r y_S^s}{y_A^a y_B^b}\right)$$

If the gases deviate from ideal conditions, the equilibrium constant for ideal gases can be used if the fugacity (f_i) is introduced in Eq. (1.67) instead of partial pressures. The relationship between these two variables and total pressure is given by:

$$f_i = \gamma_i y_i P = \gamma_i p_i$$

where γ_i is the fugacity coefficient, so that Eq. (1.67) reduces to:

$$K_e = \left(\frac{\gamma_R^r\, \gamma_S^s}{\gamma_A^a\, \gamma_B^b}\right)\left(\frac{y_R^r\, y_S^s}{y_A^a\, y_B^b}\right)P^{\Delta n}(RT)^{-\Delta n} = K_\gamma K_y\left(\frac{P}{RT}\right)^{\Delta n} \tag{1.69}$$

where:

$$K_\gamma = \left(\frac{\gamma_R^r\, \gamma_S^s}{\gamma_A^a\, \gamma_B^b}\right)$$

$$K_y = \left(\frac{y_R^r\, y_S^s}{y_A^a\, y_B^b}\right)$$

1.3 Concepts of Chemical Kinetics

1.3.1 Rate of Homogeneous Reactions

The reaction rate is a function of the conditions of the reacting system and can be affected by diverse variables. In homogeneous reaction systems, these variables are temperature, pressure and concentration, as well as the type of catalyst if the reaction is catalytic. The rate of homogeneous reaction is mostly defined as a function of the volume of the reacting system.

Considering the variation in the number of moles of the component i with respect to time, the reaction rate is defined as:

$$r_i = -\frac{1}{V}\frac{dn_i}{dt} = \frac{moles\ of\ component\ ``i"\ formed}{(volume\ of\ reacting\ system)\,(time)} \tag{1.70}$$

According to this definition, the reaction rate is equal to the rate of disappearance of the reactants and to the appearance of the products with respect to time. It will be positive for products and negative for reactants. Moreover, for the same reaction, the numerical value of the rate varies depending on what reactant or product is used, unless the stoichiometric coefficients are all the same.

For instance, for the reaction $aA + bB \rightarrow rR + sS$:

$$(-r_A) = -\frac{1}{V}\frac{dn_A}{dt} \tag{1.71}$$

$$(r_R) = \frac{1}{V}\frac{dn_R}{dt} \tag{1.72}$$

If the reacting system is at constant density, Eqs. (1.71) and (1.72) can be written as:

$$(-r_A) = -\frac{d\left(n_A/V\right)}{dt} = -\frac{dC_A}{dt} \tag{1.73}$$

$$(r_R) = \frac{d\left(n_R/V\right)}{dt} = \frac{dC_R}{dt} \tag{1.74}$$

The reaction rate also can be expressed as a function of conversion by using Eq. (1.13) in differential form:

$$dn_A = -n_{Ao}dx_A \tag{1.75}$$

Substituting Eq. (1.75) in (1.71):

$$(-r_A) = \frac{n_{Ao}}{V}\frac{dx_A}{dt} \tag{1.76}$$

If the reacting system is at constant density:

$$(-r_A) = C_{Ao}\frac{dx_A}{dt} \tag{1.77}$$

If the reacting system is at variable density, Eq. (1.76) becomes:

$$(-r_A) = \frac{n_{Ao}}{V_o(1+\varepsilon_A x_A)}\frac{dx_A}{dt} = \frac{C_{Ao}}{1+\varepsilon_A x_A}\frac{dx_A}{dt} \tag{1.78}$$

For the considered reaction, the stoichiometry shows that b moles of reactant B react each time that a moles of reactant A disappear, so that B disappears b/a times faster than A, that is:

$$(-r_B) = \frac{b}{a}(-r_A) \quad or \quad \frac{(-r_B)}{b} = \frac{(-r_A)}{a} \tag{1.79}$$

This can be generalized to any chemical species, and it is known as the *law of rates:*

$$\frac{(-r_A)}{a} = \frac{(-r_B)}{b} = \frac{(r_R)}{r} = \frac{(r_S)}{s} \tag{1.80}$$

Figure 1.3 represents Eqs. (1.73) and (1.74) in graphical form. If at any point of the curve a tangent line is plotted, its slope will be numerically

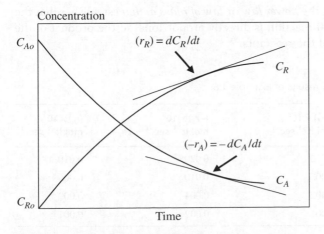

Figure 1.3 Graphical representation of the reaction rate.

equal to the reaction rate. Figure 1.3 shows that the rate changes during the reaction; at the beginning it is maximum, and it decreases when the reaction proceeds.

Example 1.8   For the gas phase reaction $A + 2B \rightarrow R$, the data reported in Table 1.5 of reaction rates at different conversions were obtained. Determine the reaction rate of all the chemical species.

Solution

From Eq. (1.80):

$$\frac{(-r_A)}{a} = \frac{(-r_B)}{b} \quad \text{and} \quad (-r_B) = \frac{b}{a}(-r_A) = \frac{2}{1}(-r_A) = 2(-r_A)$$

$$\frac{(-r_A)}{a} = \frac{(r_R)}{r} \quad \text{and} \quad (r_R) = \frac{r}{a}(-r_A) = \frac{1}{1}(-r_A) = (-r_A)$$

With the values of $(-r_A)$ given in the table and with these equations, the values of $(-r_B)$ and (r_R) can be calculated, which are summarized in Table 1.5.

1.3.2   Power Law

The rate of homogeneous reactions is only a function of the reaction conditions (pressure, temperature and concentration), that is:

$$(-r_A) = f(T,P,C_i)$$

where T, P and C_i are interdependent; that is, one of them is calculated if the other two are known, so that this equation can be rewritten as:

$$(-r_A) = f(T,C_i) = f(T)f(C_i)$$

Considering this, the *power law* or *law of mass action* establishes that the rate of a chemical reaction is directly proportional to the product of the concentrations of the reactants.

Table 1.5 Data and results of Example 1.8.

x_A	$(-r_A) \times 10^3$ mol lt^{-1} sec^{-1}	$(-r_B) \times 10^3$ mol lt^{-1} sec^{-1}	$(r_R) \times 10^3$ mol lt^{-1} sec^{-1}
0	0.010	0.020	0.010
0.2	0.005	0.010	0.005
0.4	0.002	0.004	0.002
0.6	0.001	0.002	0.001

For instance, for the reaction $aA + bB \rightarrow rR + sS$:

$$(-r_A) \propto C_A^\alpha C_B^\beta$$
$$(-r_A) = k_c C_A^\alpha C_B^\beta = k C_A^\alpha C_B^\beta \qquad (1.81)$$

where k_c or simply k is the reaction rate coefficient that only depends on temperature, and α and β are the individual reaction orders of species A and B, respectively, which can be integers or fractional.

For a gas phase system, the power law also can be expressed as a function of partial pressures:

$$(-r_A) = k_p p_A^\alpha p_B^\beta \qquad (1.82)$$

where k_p is the reaction rate coefficient as a function of partial pressures.

1.3.2.1 Relationship between k_p and k_c

For a system that follows the ideal gas law:

$$p_A = C_A RT \qquad (1.83)$$

$$p_B = C_B RT \qquad (1.84)$$

Substitution of Eqs. (1.83) and (1.84) in Eq. (1.82) gives:

$$(-r_A) = k_p (C_A RT)^\alpha (C_B RT)^\beta = k_p C_A^\alpha (RT)^\alpha C_B^\beta (RT)^\beta$$
$$(-r_A) = k_p (RT)^{\alpha+\beta} C_A^\alpha C_B^\beta = k_p (RT)^n C_A^\alpha C_B^\beta \qquad (1.85)$$

where $n = \alpha + \beta$ is the global reaction order.

The comparison of Eqs. (1.85) and (1.81) results in:

$$k = k_c = k_p (RT)^n \qquad (1.86)$$

1.3.2.2 Units of k_c and k_p

Units of k_c can be derived from Eq. (1.81):

$$k_c = \frac{(-r_A)}{C_A^\alpha C_B^\beta} [=] \frac{\left(\dfrac{mol}{lt\,sec}\right)}{\left(\dfrac{mol}{lt}\right)^\alpha \left(\dfrac{mol}{lt}\right)^\beta} [=] \frac{\left(\dfrac{mol}{lt\,sec}\right)}{\left(\dfrac{mol}{lt}\right)^{\alpha+\beta}} [=] \frac{\left(\dfrac{mol}{lt\,sec}\right)}{\left(\dfrac{mol}{lt}\right)^n} [=] \left(\dfrac{mol}{lt}\right)^{1-n} \left(\dfrac{1}{sec}\right)$$

Hence the units of k_c or k are functions of the units of concentration and time in the following way:

$$k [=] C^{1-n} t^{-1} \qquad (1.87)$$

Units of k_p can be derived from Eq. (1.82):

$$k_p[=]\frac{(-r_A)}{p_A^\alpha p_B^\beta}[=]\frac{\left(\dfrac{mol}{lt\,sec}\right)}{(atm)^\alpha(atm)^\beta}[=]\left(\frac{mol}{lt\,sec}\right)\left(\frac{1}{atm}\right)^n$$

The units of k_p are functions of the units of concentration, time and pressure:

$$k_p[=]Ct^{-1}p^{-n} \tag{1.88}$$

When the reaction rate is expressed as the variation of partial pressure with respect to time (dp_i/dt), the units of the reaction rate coefficient k_p are obtained in a similar way as k_c but with units of pressure:

$$k_p[=]p^{1-n}t^{-1} \tag{1.89}$$

It is important to note that the units of the reaction rate coefficient depend on the reaction order. For instance, for the first reaction order, k_c has units of t^{-1} (e.g. min^{-1}); and for the second reaction order, k_c has units of $C^{-1}t^{-1}$ (e.g. lt $gmol^{-1}$ min^{-1}).

Example 1.9

1) The gas phase thermal decomposition of dimethyl ether, which is carried out in a closed vessel, was studied by Hinshelwood and Askey (1927) at 504 °C. A value of $k = 4.13 \times 10^{-4}$ sec^{-1} was reported. Calculate the value of k_p.

$$(CH_3)_2O \rightarrow CH_4 + H_2 + CO \quad (A \rightarrow Products)$$

Solution

From Eq. (1.87) and the units of the reaction rate coefficient k (sec^{-1}), the following equality can be established:

$$k[=]C^{1-n}t^{-1}[=]sec^{-1}$$

This indicates that the term C^{1-n} must be equal to unity to be consistent with the units of k, so that $C^{1-n} = C^0 = 1$, and hence $1-n = 0$, therefore $n = 1$. Then, k_p can be calculated with Eq. (1.86):

$$k_p = \frac{k}{(RT)^n} = \frac{k}{RT} = \frac{4.13 \times 10^{-4}}{(0.08205)(504 + 273.15)} = 6.478 \times 10^{-6}\frac{gmol}{lt\,atm\,sec}$$

2) Kistrakowsky and Lacker (1936) studied the condensation of acrolein and butadiene at 291.2 °C, and they found a global reaction order of 2, with $k_p = 2.71 \times 10^{-11}$ $gmol$ sec^{-1} lt^{-1} $mmHg^{-2}$. Calculate k_c.

Solution

$$k = k_c = k_p (RT)^n = (2.71 \times 10^{-11}) [(62.36)(291.2 + 273.15)]^2$$

$$= 3.355 \times 10^{-2} \frac{lt}{gmol \ sec}$$

1.3.3 Elemental and Non-elemental Reactions

An *elemental reaction* is that in which the individual reaction orders are the same as the stoichiometric coefficients of the same chemical species. For instance, if in the reaction $aA + bB \rightarrow rR + sS$, $\alpha = a$ and $\beta = b$, it is an elemental reaction. When at least one of the reaction orders is not the same as the corresponding stoichiometric coefficient, the reaction is non elemental.

Example 1.10

1) *Elemental reactions*
   - The formation of hydroiodic acid is an elemental reaction because:

     $$H_2 + I_2 \rightarrow 2HI$$
     $$(-r_{H_2}) = kC_{H_2}C_{I_2}$$
     $\alpha = a = 1$ and $\beta = b = 1$.

   - The reaction between sodium hydroxide and methyl bromide is elemental because:

     $$NaOH + CH_3Br \rightarrow CH_3OH + NaBr$$
     $$(-r_{CH_3Br}) = kC_{NaOH}C_{CH_3Br}$$
     $\alpha = a = 1$ and $\beta = b = 1$.

2) *Non-elemental reactions*
   - The decomposition of nitrous oxide is not an elemental reaction because there is not agreement between the reaction order and stoichiometric coefficient:

     $$N_2O \rightarrow N_2 + \tfrac{1}{2}O_2$$
     $$(-r_{N_2O}) = \frac{k_1 C_{N_2O}^2}{1 + k_2 C_{N_2O}}$$

   - The reaction

     $$CO + Cl_2 \rightarrow COClO_2$$
     $$(-r_{CO}) = kC_{CO}C_{Cl2}^{1.5}$$

     is not elemental because $\beta \neq b$ ($\beta = 1.5$ and $b = 1$).

1.3.4 Comments on the Concepts of Molecularity and Reaction Order

As was mentioned in this chapter, molecularity is the number of molecules, atoms or ions that participate in a reaction; they can take values of 1 or 2, and sometimes 3. For instance, for the reaction $H_2 + I_2 \rightarrow 2HI$, the molecularity is $a + b = 1 + 1 = 2$.

Elemental reactions with molecularity of 3 have not been found so far; thus, the reaction of ammonia formation ($N_2 + 3H_2 \rightarrow 2NH_3$) cannot be elemental.

Non-elemental reactions can be explained by assuming that what is observed as a simple reaction is, in fact, the global effect of a sequence of elemental reactions in which the intermediate products are negligible and are not detected (Butt, 1980).

The *reaction order* is defined as the exponent to which the concentrations or partial pressures are raised in the kinetic expression. For instance, in Eqs. (1.81) and (1.82), α is the reaction order with respect to A and β is the reaction order with respect to B.

The global reaction order (n) is defined as the sum of the individual reaction orders. For the previous case, $n = \alpha + \beta$. In the case of the non-elemental decomposition of nitrous oxide, it makes no sense to talk about reaction order.

It should be remembered that because kinetic equations are determined from experimental data, the exponents are sometimes fractional, and in such cases the reaction order is also fractional. Table 1.6 shows some examples of chemical reactions with different orders.

1.3.5 Dependency of k with Temperature

1.3.5.1 Arrhenius Equation

Temperature has an important influence on the rate of chemical reactions. In 1889, Arrhenius (1889) explained this dependency by means

Table 1.6 Chemical reactions with differents orders.

Reaction	Kinetic model	Global reaction order	Type of reaction
$N_2O_5 \rightarrow N_2O_4 + \frac{1}{2}O_2$	$(-r_A) = k\,C_A$	1	Elemental
$2NO + O_2 \rightarrow 2NO_2$	$(-r_A) = k\,C_A^2 C_B$	3	Elemental
$CH_3CHO \rightarrow CH_4 + CO$	$(-r_A) = k\,C_A^{1.5}$	1.5	Non-elemental
$4PH_3 \rightarrow P_4 + 6H_2$	$(-r_A) = k\,C_A$	1	Non-elemental
$CO + Cl_2 \rightarrow COCl_2$	$(-r_A) = k\,C_A\,C_B^{1/2}$	1.5	Non-elemental

of a simple exponential form based on thermodynamic considerations with the equation of Van't Hoff:

$$\frac{d\ln K_e}{dT} = \frac{\Delta H^o}{RT^2} \tag{1.90}$$

where K_e is the equilibrium constant, T the absolute temperature and ΔH^o the change of enthalpy of the reaction. Substituting Eq. (1.66) in Eq. (1.90):

$$\frac{d\ln\left(k_1/k_2\right)}{dT} = \frac{\Delta H^o}{RT^2}$$

$$\frac{d\ln k_1}{dT} - \frac{d\ln k_2}{dT} = \frac{\Delta H^o}{RT^2}$$

Arrhenius suggested that ΔH^o can be divided in two components, E_{A1} and E_{A2}, in such a way that $\Delta H^o = E_{A1} - E_{A2}$. Then, the previous equation can be separated in two equations, one for the forward reaction and another one for the reverse reaction:

$$\frac{d\ln k_1}{dT} = \frac{E_{A1}}{RT^2}$$

$$\frac{d\ln k_2}{dT} = \frac{E_{A2}}{RT^2}$$

Integration of any of these two equations, assuming that E_{A1} and E_{A2} are independent on temperature, gives:

$$\ln k = -\frac{E_A}{RT} + C_I$$

where C_I is an integration constant that can be substituted by $\ln e^{C_I} = \ln A$, where A is another constant equal to e^{C_I}. With these changes, the following operations with logarithms can be done:

$$\ln k = -\frac{E_A}{RT} + \ln e^{C_I} = -\frac{E_A}{RT} + \ln A$$

$$\ln k - \ln A = \ln\frac{k}{A} = -\frac{E_A}{RT}$$

And, finally, the Arrhenius equation is obtained:

$$k = A\ EXP(-E_A/RT) \tag{1.91}$$

where A is the frequency or pre-exponential factor, E_A the activation energy, R the universal gas constant and T the absolute temperature.

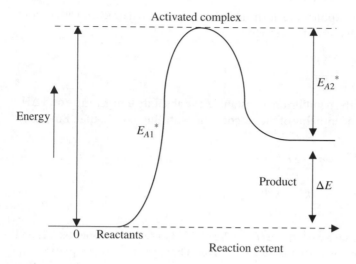

Figure 1.4 Graphical representation of the activation energy.

1.3.5.2 Frequency Factor and Activation Energy

In the Arrhenius equation, A represents the probability of effective collision of reactants to form products, and E_A is the least amount of energy required to activate atoms or molecules of reactants to a state in which they can undergo a chemical reaction to form products.

The forward reaction involves an energy E_{A1}, while the reverse reaction involves an energy E_{A2}, so that there is a difference of energy ΔE_A.

These conditions are satisfied only if the reaction proceeds by means of an intermediate state that had an energy E_{A1}^* higher than the initial state, and an energy E_{A2}^* higher than the final state.

The intermediate state is known as the *activated complex*. The molecules of reactant have to reach the energy E_{A1}^* before they can form the complex, and hence the products; this energy is called *activation energy*. Figure 1.4 illustrates this for the case of an endothermic reaction.

1.3.5.3 Evaluation of the Parameters of the Arrhenius Equation

The evaluation of the Arrhenius equation parameters requires data of k at different temperatures, and it can be done by different means depending on the manner that Eq. (1.91) is transformed into a linear equation. The following are three approaches to do this linearization.

1.3.5.3.1 Traditional Method (TM)

To evaluate A and E_A, it is necessary to transform Eq. (1.91) in a linear form by means of logarithms:

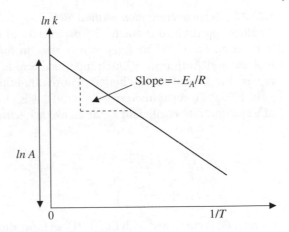

$$\ln k = \ln A - \frac{E_A}{R}\left(\frac{1}{T}\right) \tag{1.92}$$

When plotting $1/T$ versus $\ln k$, the slope of the straight line gives the value of E_A and with the intercept the value of A is obtained, as illustrated in Figure 1.5.

If only two values of T and k are available, the following equations can be derived from Eq. (1.91) to calculate E_A:

$$\ln\left(\frac{k_2}{k_1}\right) = -\frac{E_A}{R}\left(\frac{1}{T_2} - \frac{1}{T_1}\right) \tag{1.93}$$

$$E_A = \frac{RT_1 T_2}{T_2 - T_1}\ln\left(\frac{k_2}{k_1}\right) \tag{1.94}$$

In some cases, the experimental data of reaction rate coefficients with respect to temperature do not follow a linear behaviour, according to Figure 1.5. This is attributed to the following reasons (Holland and Anthony, 1979):

1) The reaction mechanism changes in the range of studied temperature.
2) The form of the rate expression does not correspond to the studied reaction.
3) Other phenomena (e.g. diffusion) that control the reaction are sufficiently slow.
4) The dependency of the frequency factor on temperature starts to become important.

1.3.5.3.2 Reparameterization Method (RM)

Another approach to evaluate the parameters of the Arrhenius equation is by using Eq. (1.91) in a reparameterization form (Himmelblau, 1970; Holland and Anthony, 1979; Drapper and Smith, 1981), which has been reported to be statistically better than the traditional method (Chen and Aris, 1992). To reparameterize Eq. (1.91), it is considered the evaluation of a reaction rate coefficient k_m at an average temperature T_m, as follows:

$$T_m = \frac{\sum_{i=1}^{n} T_i}{n} = \frac{T_1 + T_2 + \dots + T_n}{n} \tag{1.95}$$

$$k_m = A \, EXP\left(-E_A / RT_m\right) \tag{1.96}$$

If Eq. (1.96) is combined with Eq. (1.91), an expression similar to Eq. (1.92) is obtained, which is the reparameterized form of the Arrhenius equation:

$$k = k_m EXP\left[-\frac{E_A}{R}\left(\frac{1}{T} - \frac{1}{T_m}\right)\right]$$

Its linear form is:

$$\ln k = \ln k_m - \frac{E_A}{R}\left(\frac{1}{T} - \frac{1}{T_m}\right) \tag{1.97}$$

By plotting $(1/T - 1/T_m)$ versus $\ln k$, the slope of the straight line gives the value of E_A, and with the intercept the value of k_m is obtained at temperature T_m, as illustrated in Figure 1.6. A is then calculated from Eq. (1.96):

$$A = \frac{k_m}{EXP\left(-E_A / RT_m\right)} = k_m EXP\left(E_A / RT_m\right) \tag{1.98}$$

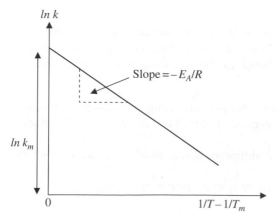

Figure 1.6 Graphical representation of the reparameterized Arrhenius equation.

1.3.5.3.3 Reduction of Orders of Magnitude (ROM)

A frequently found problem when calculating the parameters of the Arrhenius equation is that A and E_A have different orders of magnitude – 10^7 to 10^{14} and 10^3 to 10^4, respectively – and when applying the method of linear regression, the calculation is done with small differences of large values, which causes rounding error (Butt, 1980). This situation can be avoided by reducing the orders of magnitude of the parameters in the following way:

$$a_r = lnA$$

$$b_r = E_A/1000$$

$$t_r = RT/1000$$

Then, Eq. (1.91) in linear form becomes:

$$\ln k = a_r - b_r \left(\frac{1}{t_r}\right) \tag{1.99}$$

The representation of Eq. (1.99) is given in Figure 1.7.

To illustrate the methods for determining the parameters of the Arrhenius equation, two examples are presented. Example 1.10 shows the use of linear regression, and Example 1.11 presents the use of non-linear regression and a comparison of the three methods described above.

Example 1.11 Worsfold and Bywater (1960) studied the isomerization of styrene using benzene as solvent, and reported the experimental data of Table 1.7. Calculate the activation energy and the frequency factor with the traditional and reparameterization methods.

Figure 1.7 Graphical representation of the Arrhenius equation with the method of reduction of orders of magnitude.

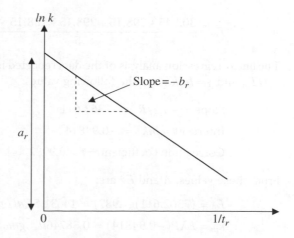

Table 1.7 Data and results of Example 1.11.

T (K)	k ($lt^{0.5}$ gmol$^{0.5}$ min^{-1})	$\ln k$	$1/T \times 10^3$	$1/T - 1/T_m$
303.45	0.929	−0.07365	3.29544	-1.15787×10^{-4}
298.15	0.563	−0.57448	3.35402	-5.72065×10^{-5}
293.15	0.387	−0.94933	3.41122	0
288.15	0.265	−1.32803	3.47041	5.91918×10^{-5}
283.15	0.155	−1.86433	3.53170	1.20474×10^{-4}

Solution

- *Traditional method*: The linear regression analysis of the data reported in Table 1.7 for $x = 1/T$ and $y = \ln k$ gives the following values:

$$\text{Slope} = -E_A/R = -7363.608$$
$$\text{Intercept} = \ln A = 29.310$$
$$\text{Correlation coefficient} = r = 0.9982$$

From these values, A and E_A are:

$$E_A = -(-7363.608)(1.987) = 14631.5 Cal/gmol$$
$$A = EXP(29.310) = 3.1422 \times 10^{10} lt^{0.5} gmol^{0.5} min^{-1}$$

- *Reparameterization method*: The average temperature is:

$$T_m = \frac{303.45 + 298.15 + 293.15 + 288.15 + 283.15}{5} = 293.15 K$$

The linear regression analysis of the data reported in Table 1.7 for $x = 1/T - 1/T_m$ and $y = \ln k$ gives the following values:

$$\text{Slope} = -E_A/R = -7363.611$$
$$\text{Intercept} = \ln k_m = -0.94814$$
$$\text{Correlation coefficient} = r = 0.9982$$

From these values, A and E_A are:

$$E_A = (7362.611)(1.987) = 14631.5 Cal/gmol$$
$$k_m = EXP(-0.94814) = 0.38746 lt^{0.5} gmol^{0.5} min^{-1}$$

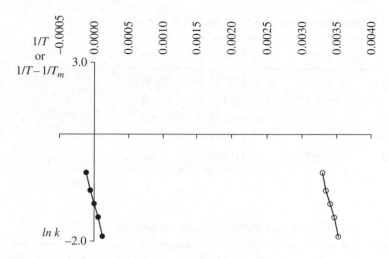

Figure 1.8 Results of example 1.11. (○) TM, (●) RM.

A is obtained from Eq. (1.98):

$$A = k_m EXP\left(\frac{E_A}{RT_m}\right) = (0.38746)EXP\left(\frac{14631.5}{(1.987)(293.15)}\right)$$
$$= 3.1422 \times 10^{10} lt^{0.5} gmol^{0.5} min^{-1}$$

Figure 1.8 shows the results of the linear regression with the two approaches, so that the final Arrhenius equation is:

$$k = 3.1422 \times 10^{10} \, EXP\left(-\frac{14631.5}{RT}\right)$$

The values of parameters of the Arrhenius equation for this example had the same results; however, it has been reported that when large numbers of experimental data in a wide range of temperature values are used, the reparameterized methods give better results.

Example 1.12   To illustrate the way in which the parameters of the Arrhenius equation are calculated with the three approaches described above using non-linear regression analysis, three cases of kinetic studies reported in the literature are considered.

Case 1. Isomerization of styrene. The same data as in Example 1.11
Case 2. Thermal cracking of ethane (Froment *et al.*, 2010). The experimental data are reported in Table 1.8.

Table 1.8 Experimental data of thermal cracking of ethane.

T (K)	k (sec^{-1})	T (K)	k (sec^{-1})
975.15	0.150	1062.15	1.492
998.15	0.274	1076.15	2.138
1007.15	0.333	1083.15	2.718
1027.15	0.595	1100.15	4.317
1046.15	0.923	1110.15	4.665

Table 1.9 Experimental data of the hydrogenation of ethylene.

T (K)	k (gmol/cm^{3}sec)	T (K)	k (gmol/cm^{3}sec)	T (K)	k (gmol/cm^{3}sec)
350.15	281	326.05	70.4	327.65	70
350.15	288	350.7	240	283.15	2.57
336.65	148.5	335.9	142	283.15	1.99
326.45	71	326.9	69.1	285.5	4.11
326.45	66.5	326.9	67.8	285.5	3.79
350.8	244	352.65	303	285.5	3.84
350.8	240	352.65	306	285.5	3.90
350.8	126.5	337.1	131	285.5	3.88
326.05	72.3	337.1	136.9	285.5	3.74

Case 3. Hydrogenation of ethylene (Wynkoop and Wilhelm, 1950). The experimental data are reported in Table 1.9.

Solution

The calculation of the parameters of the Arrhenius equation by non-linear regression analysis requires the definition of an objective function to be minimized, which is commonly given by the sum of square errors (*SSE*) between experimental and calculated values as follows:

$$SSE = \sum_{i=1}^{n} \left(\ln k_i^{exp} - \ln k_i^{calc} \right)^2 = \sum_{i=1}^{n} \left(\ln \frac{k_i^{exp}}{k_i^{calc}} \right)^2$$

Using this equation for the three approaches results in:

Traditional Method (Eq. 1.92).

$$SSE_{TM} = \sum_{i=1}^{n} \left(\ln k_i^{exp} - \ln A + \frac{E_A}{RT} \right)^2$$

Reparameterization Method (Eq. 1.97).

$$SSE_{RM} = \sum_{i=1}^{n} \left(\ln k_i^{exp} - \ln k_m + \frac{E_A}{RT} \left(\frac{1}{T} - \frac{1}{T_m} \right) \right)^2$$

Reduction of Orders of Magnitude (Eq. 1.99).

$$SSE_{ROM} = \sum_{i=1}^{n} \left(\ln k_i^{exp} - a_r + \frac{b_r}{t_r} \right)^2$$

By minimizing these objective functions, the values reported in Table 1.10 were obtained.

Table 1.10 Results of the non-linear regression of Example 1.12.

	TM	RM	ROM
Case 1			
A (lt$^{0.5}$ gmol$^{0.5}$ min^{-1})	3.134×10^{10}	3.142×10^{10}	3.141×10^{10}
E_A (Cal/mol)	14631	14631	14631
r	0.9982	0.9982	0.9982
SSE	0.0067	0.0067	0.0067
Case 2			
A (gmol/cm^{3}sec)	7.541×10^{11}	7.570×10^{11}	7.558×10^{11}
E_A (Cal/mol)	56827	56835	56831
r	0.9985	0.9991	0.9989
SSE	0.0364	0.0364	0.0364
Case 3			
A (sec^{-1})	5848	5844	5839
E_A (Cal/mol)	13329	13329	13329
r	0.9955	0.9955	0.9955
SSE	0.7676	0.7676	0.7676

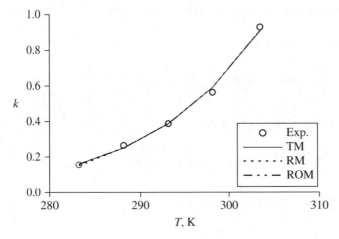

Figure 1.9 Results of Case 1 for example 1.12.

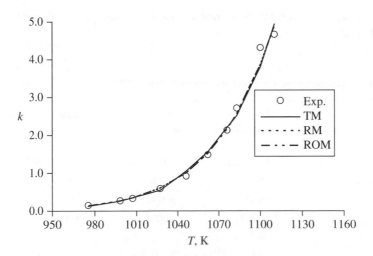

Figure 1.10 Results of Case 2 for example 1.12.

Figures 1.9 and 1.10 show a comparison of experimental data and those calculated with the three approaches for Cases 1 and 2, respectively. It is seen that, apparently, there are not significant differences in the predictions. The same happens for Case 3, which is why Figure 1.11 only shows the prediction with the traditional method.

For Cases 1 and 3, the calculated activation energies were the same when using the three methods (14,631 and 13,329 Cal/mol, respectively). However, for Case 2, the values of E_A were in the range of 56,827–56,835 Cal/mol; although the difference is small, the wider range of reaction

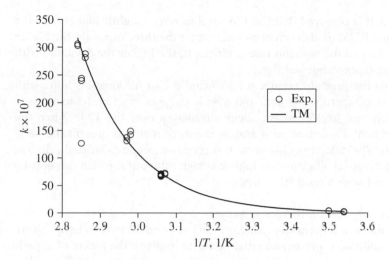

Figure 1.11 Results of Case 3 for example 1.12.

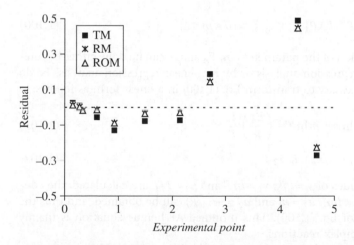

Figure 1.12 Residuals analysis for Case 2 for example 1.12.

temperatures (20 K, 135 K and 67 K for Cases 1, 2 and 3, respectively) starts to have a small effect on the estimation of the parameters, which according to literature reports (Himmelblau, 1970; Drapper and Smith, 1981) causes higher error than the traditional method. This was also slightly observed in the values of the correlation coefficient.

The range of the values of the reaction rate coefficient does not have a significant effect on the estimation of the parameters, since for Case 3, which has the wider range, the prediction was practically the same.

On the other hand, Figure 1.12 shows the residuals (differences between the experimental and calculated values of reaction rate coefficients) for

Case 2. It is observed that the TM exhibits errors slightly higher than the RM and ROM. It also can be noted that for the three methods, the higher the values of the reaction rate coefficients, the higher the deviation with respect to experimental data.

From the previous results, it is concluded that for kinetic studies with sufficient experimental data and a wide range of reaction temperatures, the RM and ROM present slight advantages over the TM. When the experimental information is scarce, the three methods present the same results. The differences between the experimental and calculated values of reaction rate coefficients are higher at high values of k, which can only be observed with a residual analysis.

1.3.5.4 Modified Arrhenius Equation

Sometimes, it is necessary to include an independent term of temperature as an additional pre-exponential factor to improve the fitting of experimental data of k versus T. This is usually done by means of T raised to a power, as follows:

$$k = AT^m \; EXP\left(-E_A/RT\right) \quad (0 \leq m \leq 1) \tag{1.100}$$

The calculation of the parameters m, E_A and A can be done either by multiple linear regression analysis or by non-linear regression analysis. To do this, it is necessary to transform Eq. (1.100) in a linear form as follows:

$$\ln k = \ln A + m\ln T - \left(\frac{E_A}{R}\right)\frac{1}{T} \tag{1.101}$$

$$y = a_o + a_1 x_1 + a_2 x_2 \tag{1.102}$$

Once the values of $y = \ln k$, $x_1 = \ln T$ and $x_2 = 1/T$ are calculated, the constants a_i ($a_o = \ln A$, $a_1 = m$ and $a_2 = -E_A/R$) can be obtained, and thus the parameters of Eq. (1.100). This modified Arrhenius equation is mainly used for complex reactions.

There are other theories to represent the dependency of the reaction rate coefficient with temperature, which are specific cases of Eq. (1.100), among them are the following:

1) *Theory of collision.* It considers that the reaction rate depends on the number of energetic collisions between reactants no matter what happens with unstable intermediate product, by considering that the reactant decomposes so rapidly in products that it does not influence the global reaction rate.

$$k = AT^{1/2} \; EXP\left(-E_A/RT\right) \tag{1.103}$$

In linear form:

$$\ln\frac{k}{T^{1/2}} = \ln A - \left(\frac{E_A}{R}\right)\frac{1}{T}$$

$$y = \ln\frac{k}{T^{1/2}} \quad x = \frac{1}{T}$$

(1.104)

2) *Theory of the steady state.* It considers that the reaction rate depends on the rate of decomposition of the intermediate product. It assumes that the rate of the intermediate product formation is so fast that at any moment, its concentration is that of the equilibrium, no matter how it can be formed.

$$k = AT \; EXP\left(^{-E_A}/_{RT}\right)$$

(1.105)

In linear form:

$$\ln\frac{k}{T} = \ln A - \left(\frac{E_A}{R}\right)\frac{1}{T}$$

$$y = \ln\frac{k}{T} \quad x = \frac{1}{T}$$

(1.106)

1.4 Description of Ideal Reactors

Chemical reactors can have a great variety of sizes, modes of operation and operating conditions. However, to generate kinetic data, there are two main types of reactors: batch reactors and continuous reactors. The batch reactor is the most commonly used for conducting experiments that aim at obtaining the variation of any property by means of which conversion can be calculated as a function of reaction time. Due to this, in this section and in the whole book, more emphasis is placed on the batch reactor.

1.4.1 Batch Reactors

Batch or discontinuous reactors used for kinetic measurements are generally a completely sealed pressure vessel with an agitator, an external electrical heating system, an internal cooling system, pressure and temperature sensors and chemical analysis equipment incorporated with an overpressure safety system. A typical diagram of a laboratory installation of a batch reactor is shown in Figure 1.13.

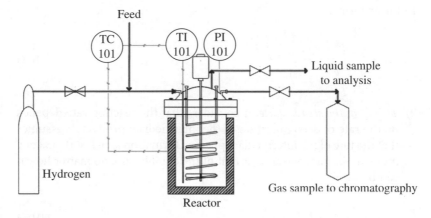

Figure 1.13 Typical experimental setup of a batch reactor.

The internal cooling system allows for better temperature stability when reactions are exothermic. Additionally, the agitation system provides an efficient mixing within the discontinuous reactor, which ensures homogeneous temperature and composition gradients of reactants and products at any time (Angeles *et al.*, 2014).

This type of reactor is used not only for the study of reaction kinetics, but also for conducting studies in early stages of process development such as determination of optimal operating conditions and screening of catalysts. Batch reactors are easy to operate because the reactants and catalyst (if present) are loaded to the recipient, and the products can be discharged and analysed at any desired time during the reaction (Pitault *et al.*, 2004).

The volume of the reactors used in the laboratory lies between 50 and 3000 mL. The size of the reactor limits the mode of operation and the number of samples that can be taken and analysed as a function of reaction time.

1.4.1.1 Modes of Operation
There are two modes of operation in a batch reactor aiming at determining the reaction kinetics: isothermal and temperature scanning.

Regardless of the method of operation selected, the batch operation is carried out once the reaction pressure is achieved. During the experiment, the reactants are not added (batch operation). A common practice is to keep the initial pressure, temperature and amount of reactants constant in all tests while the effect of time is evaluated.

Before any kinetic study, preliminary tests to determine the operating conditions (pressure, temperature, time, stirring rate, etc.) to work under the kinetic regime need to be conducted. It is important to ensure that there are no temperature or concentration gradients within the reactor. Therefore, proper selection of stirring speed allows for a homogeneous reaction mixture (Perego and Peratello, 1999).

1.4.1.1.1 Isothermal Operation

In isothermal operation, the collection of kinetic data consists in determining how the composition of the reaction mixture varies with time at defined operating conditions. Several tests are performed by varying only one parameter at a time. The initial reaction time is set once the reactor is heated for a period of time without any stirring until the desired reaction temperature and pressure are reached (Nguyen *et al.*, 2013).

Operating temperature remains constant, and throughout the test, a small amount of samples might be taken for analysis at regular time intervals. Once the reaction time finishes, the reactor is discharged, and the final products are collected and analysed.

During the heating time necessary to achieve and stabilize the temperature of the reactor, useless data are obtained even though thermal reactions are occurring which are not proceeding at constant operating conditions. Useful samples can be collected only immediately after the heating period, so missing data during preheating may affect the determination of reaction rates (Helfferich, 2003).

Nevertheless, isothermal operation of a batch reactor can still be possible at different reaction times, holding constant the other operating conditions. This type of operation has the advantage that the products obtained at each reaction time correspond to only a single reaction temperature. The analysis of the products gives additional information about the effects of operating conditions and possible reaction mechanisms.

Moreover, a small amount and number of samples are taken in the isothermal operation, which makes the reactors used in this type of operation smaller and easy to handle in the laboratory (Rezaei *et al.*, 2010).

1.4.1.1.2 Temperature Scanning Operation

In contrast to isothermal operation, the scanning operation of a batch reactor allows deliberately for varying the temperature during a test. The heating–cooling systems of the reactor and the heat of reaction are used to reach another reaction temperature. Thus, the heating rate can be controlled to bring the reactor to another preselected reaction temperature without affecting significantly the calculation of the reaction rates caused by the undesirable thermal reactions (Wojciechowski, 1997).

The temperature scanning operation is always more productive than isothermal operation in a batch reactor for the following reasons (Rice, 1997):

1) It does not need to achieve a constant temperature and discard all data collected before an operating condition is reached. It is possible to operate at a lower temperature before reaching the desired reaction temperature, and thus to study the thermal cracking reactions.
2) It has complete control over the heating rate by varying the temperature ramps, which allows for obtaining different conversion trajectories versus time.
3) Conversion curves generated at different heating rates allow for obtaining more accurate reaction rates.
4) It reduces the number of experiments because changes in temperature in the batch reactor permit the simultaneous study of various reaction pressures.

However, the temperature scanning method requires using a larger volume of reactor to avoid any disturbance of the batch operation. The experiment is carried out at two or more reaction temperatures during the same test, then a larger number of samples are taken to characterize the products.

1.4.1.2 Data Collection

During the study of reaction kinetics in batch reactors, the following experimental data are typically collected.

1.4.1.2.1 *Pressure and Temperature inside the Reactor*

During isothermal operation of a batch reactor, once reaction temperature is reached, agitation is initiated, and this moment is assumed to be the beginning of the reaction (i.e. the reaction time equals zero). When the agitation in the reactor is switched at the beginning of the test, a small pressure drop can be observed.

The experimental data collection is done at regular intervals, generating profiles of pressure and temperature of the reactor such as those shown in Figure 1.14. This figure shows the case of a gas phase reaction in which the reaction pressure decreases with time (Nguyen *et al.*, 2013).

In the case of a temperature scanning operation, several experiments at different temperature and pressure are performed. This allows not only for obtaining profiles as a function of time (similar to those shown in Figure 1.15), but also for reducing the number of experiments (Matsumura, 2005).

1.4.1.2.2 *Properties and Composition of the Products*

To determine the reaction kinetics, the most common procedure is to measure the properties and composition of products versus time for a set of operating conditions. For any selected operation method, the

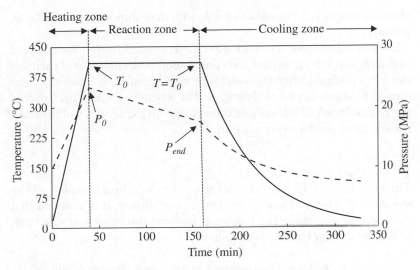

Figure 1.14 Typical pressure (---) and temperature (—) profiles in a batch reactor operated isothermally.

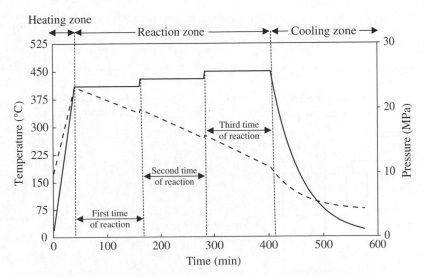

Figure 1.15 Typical pressure (---) and temperature (—) profiles in a batch reactor operated by temperature scanning.

chemical kinetics can be obtained only with the composition of the liquid product.

Samples taken from the reactor at different time intervals are analysed to determine their properties and composition. Mass balances of each test can be performed after the reactor is discharged. The amount of the formed products can be easily obtained by weighting. Individual balances can be developed which provide information on the accuracy of the measurements of product composition.

1.4.1.3 Mass Balance

The mass balance in a batch reactor can be done by examining all the streams indicated in Figure 1.16. Due to the stirring, it is assumed that the composition inside the reactor is uniform throughout at any time. The general mass balance is as follows:

$$Inlet = Outlet + Disapearance\,by\,reaction + Accumulation \tag{1.107}$$

Since during reaction, neither reactants are fed nor products are withdrawn, the inlet and outlet are equal to zero. Thus, Eq. (1.107) is:

$$Disapearance\,by\,reaction = -Accumulation \tag{1.108}$$

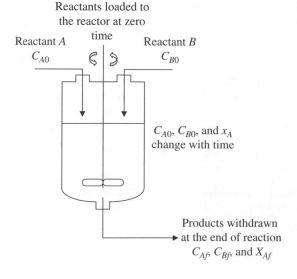

Reactants loaded to the reactor at zero time

Reactant A C_{A0}

Reactant B C_{B0}

C_{A0}, C_{B0}, and x_A change with time

Products withdrawn at the end of reaction C_{Af}, C_{Bf} and X_{Af}

Figure 1.16 Schematic of a batch reactor.

This equation can be expressed in terms of the limiting reactant A:

$$Rate \ of \ disapearance \ of \ A \ by \ reaction = -Rate \ of \ accumulation \ of \ A$$

(1.109)

The left-hand side of Eq. (1.109) is:

$$Rate \ of \ disapearance \ of \ A \ by \ reaction = (-r_A)V$$

$$= \frac{moles \ of \ A \ that \ react}{(time)(volume)}(volume)$$

where V is the volume of reacting fluid. It must be noted that V is not necessarily equal to the reactor volume.

The right-hand side of Eq. (1.109) is:

$$Rate \ of \ accumulation \ of \ A = \frac{dn_A}{dt} = -n_{Ao}\frac{dx_A}{dt}$$

Substituting both terms in Eq. (1.108):

$$-\frac{dn_A}{dt} = n_{Ao}\frac{dx_A}{dt} = (-r_A)V$$

(1.110)

Eq. (1.110) is the general equation of design of batch reactors in differential form, which can be integrated for the cases of constant or variable volume of the reacting fluid. This equation is used in the remaining chapters of this book for deducing the different kinetic models depending on the type of reaction, type of feed and so on.

The typical form of Eq. (1.110) is:

$$t = n_{Ao} \int_0^{x_A} \frac{dx_A}{(-r_A)V}$$

(1.111)

1.4.2 Continuous Reactors

Continuous chemical reactors are used mainly for studying the kinetics of heterogeneous chemical reactions; however, they also are used sometimes for determining the kinetics of homogeneous reactions. There are two types of continuous reactors: (1) PFR: a plug flow reactor or tubular reactor, in which the flow of reactants through it is orderly, and no element of fluid is mixed with any other element ahead or behind; and (2) CSTR: a continuous stirred tank reactor or back-mix reactor, characterized by its content, is perfectly mixed, so that the composition at the exit is the same as that inside the reactor. The CSTR and PFR are the two ideal limits of mixing: completely mixed and not mixed at all, respectively. All real flow reactors will lie somewhere between these two limits.

The generation of kinetic data in the continuous reactors is done mainly by measuring the concentration or conversion as a function of the feed flowrate at constant pressure and temperature.

1.4.2.1 Space–Time and Space–Velocity

Differently from batch reactors, in which concentration or conversion change with respect to time, in continuous reactors the concentration is the same at any time if the reaction conditions are constant. Instead of time, in continuous reactors there is a need for another parameter by which the changes in concentration can be followed. These parameters are space–time or space–velocity, which are defined as follows: *space–time* is the time required to process one reactor volume of feed, while *space–velocity* is the number of reactor volumes of feed which can be treated in unit time. Both are typically measured at the conditions of the feed entering the reactor. Space–time (τ) and space–velocity (SV) are calculated as:

$$\tau = \frac{C_{A0} V}{F_{A0}} = \frac{V}{v_0} [=] time \tag{1.112}$$

$$SV = \frac{1}{\tau} = \frac{v_0}{V} [=] time^{-1} \tag{1.113}$$

where F_{Ao} and v_o are the molar and volumetric flowrates of A, respectively.

1.4.2.2 Plug Flow Reactor

The PFR behaves as an ideal reactor when the reactants move in the axial direction without mixing with any element ahead or behind, and the residence time in the reactor is the same for all the elements of the fluid. It is assumed that no mixing occurs between adjacent fluid volume elements either radially or axially. When the operation reaches the steady state, any property of the system varies with respect to time in a certain point of the reactor. Temperature, pressure and composition can vary with respect to residence time or reactor length.

The mass balance in a PFR can be done by examining all the streams indicated in Figure 1.17 as follows:

Inlet = Outlet + Dissapearance by reaction + Accumulation (1.114)

At steady-state conditions, there is no accumulation:

Inlet = Outlet + Disapearance by reaction (1.115)

Because the properties of the reacting system change with the reactor length, the mass balance is done in differential form. For the differential

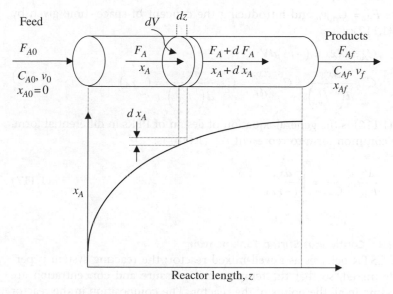

Figure 1.17 Schematic of a plug flow reactor.

element of volume shown in Figure 1.17, each term of Eq. (1.115) for the limiting reactant A is:

Rate of inlet of A = F_A

Rate of outlet of A = $F_A + dF_A$

Rate of disapearance of A by reaction $= (-r_A)dV$

Substituting in Eq. (1.115):

$$F_A = F_A + dF_A + (-r_A)dV$$

$$-dF_A = (-r_A)dV$$

Since $F_A = F_{Ao}(1-x_A)$:

$$dF_A = -F_{Ao}dx_A$$

so that:

$$F_{Ao}dx_A = (-r_A)dV$$

Since $F_{Ao} = C_{Ao}v_o$ and introducing the concept of space–time given by Eq. (1.112):

$$C_{Ao}v_o dx_A = (-r_A)dV$$

$$C_{Ao}\frac{dx_A}{d\left(V/_{v_o}\right)} = C_{Ao}\frac{dx_A}{d\tau} = C_{Ao}\frac{dx_A}{d\left(1/_{SV}\right)} = (-r_A) \tag{1.116}$$

Eq. (1.116) is the general equation of design of PFRs in differential form. The common form to represent Eq. (1.116) is:

$$\frac{V}{F_{Ao}} = \frac{\tau}{C_{Ao}} = \int_0^{x_A}\frac{dx_A}{(-r_A)} \tag{1.117}$$

1.4.2.3 Continuous Stirred Tank Reactor

The CSTR behaves as a well-mixed reactor; the reacting system is perfectly mixed, so that the temperature, pressure and concentration are the same in all the points of the reactor. The composition in the reactor is assumed to be that of the effluent stream, and therefore all the reaction occurs at this constant composition.

The mass balance in a CSTR can be done by examining all the streams indicated in Figure 1.18, as follows:

$$Inlet = Outlet + Disapearance\ by\ reaction + Accumulation$$
$$\tag{1.118}$$

At steady-state conditions, there is no accumulation:

$$Inlet = Outlet + Disapearance\ by\ reaction \tag{1.119}$$

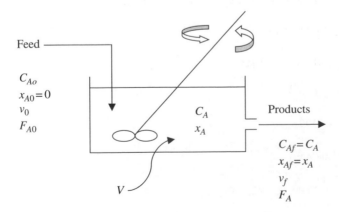

Figure 1.18 Schematic of a continuous stirred tank reactor.

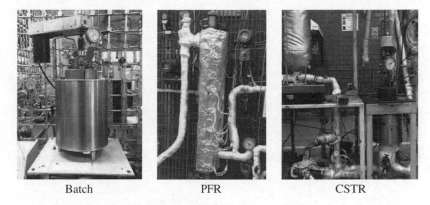

Batch   PFR   CSTR

Figure 1.19 Examples of a laboratory batch reactor, PFR and CSTR.

Because the properties of the reacting system are the same in the whole reactor, the mass balance is done for the entire volume of the system. Each term of Eq. (1.119) for the limiting reactant A is:

$$Rate\ of\ inlet\ of\ A = F_{Ao}$$

$$Rate\ of\ outlet\ of\ A = F_A$$

$$Rate\ of\ disapearance\ of\ A\ by\ reaction = (-r_A)V$$

Substituting in Eq. (1.115):

$$F_{Ao} = F_A + (-r_A)V$$

Since $F_A = F_{Ao}(1 - x_A)$:

$$F_{Ao}x_A = (-r_A)V$$

Since $F_{Ao} = C_{Ao}v_o$ and introducing the concept of space–time given by Eq. (1.112):

$$C_{Ao}v_o x_A = (-r_A)V$$

$$C_{Ao}\frac{x_A}{\left(V/v_o\right)} = C_{Ao}\frac{x_A}{\tau} = C_{Ao}\frac{x_A}{1/sv} = (-r_A) \tag{1.120}$$

Eq. (1.120) is the general equation of design of CSTRs in differential form. Another form of Eq. (1.119) is:

$$\frac{V}{F_{Ao}} = \frac{\tau}{C_{Ao}} = \frac{x_A}{(-r_A)} \tag{1.121}$$

Figure 1.19 shows examples of a batch reactor, PFR and CSTR typically used in experimental studies.

2

Irreversible Reactions of One Component

Irreversible reactions involving only one component have the following general form:

$$aA \rightarrow Products \tag{2.1}$$

For the power-law model, they have the following kinetic expression:

$$(-r_A) = kC_A^n \tag{2.2}$$

If $n = a$, the reaction in Eq. (2.1) is elemental, while if $n \neq a$, the reaction is not elemental. Some examples of elemental and non-elemental irreversible reactions with one component are presented in Table 2.1.

The analysis of the kinetic data for irreversible reactions with one component can be done by the following methods:

- *Integral method*: A reaction order is assumed ($0 < n < 3$), the corresponding power-law kinetic model is integrated and the resulting expression is transformed in a straight-line equation which should have two terms: one term containing the time and the reaction rate coefficient, and another term with a function of concentration. If, in a graphical representation of the two terms, the experimental data fit a straight line, the reaction rate coefficient is calculated with the value of the slope (graphical method) or by linear regression analysis. The reaction rate coefficient can also be determined from the assumed kinetic model by calculating its different values for each pair of experimental points (substitution or analytical methods). If the reaction rate coefficient shows similar values without any tendency, the proposed kinetic model is correct.
- *Differential method*: Eq. (2.2) is used directly in differential form without integration, and by means of logarithms it is transformed into a straight-line equation. The reaction rate is then evaluated with the experimental data. The reaction order and the reaction rate coefficient are determined by the resulting linear expression.

Chemical Reaction Kinetics: Concepts, Methods and Case Studies, First Edition.
Jorge Ancheyta.
© 2017 John Wiley & Sons Ltd. Published 2017 by John Wiley & Sons Ltd.

Table 2.1 Examples of irreversible reactions with one component.

Reaction	Kinetic model	Temperature
$CH_3OCH_3 \rightarrow CH_4 + CO + H_2$	$(-r_A) = 4.94 \times 10^{-3}(\text{sec}^{-1})C_A$	552 °C
$2AsH_{3\,(g)} \rightarrow As_{2\,(s)} + 3H_{2\,(g)}$	$(-r_A) = 1.52 \times 10^{-2}(h^{-1})C_A$	350 °C
$N_2O_5 \rightarrow N_2O_4 + \frac{1}{2}O_2$	$(-r_A) = 2.7 \times 10^{-4}(\text{sec}^{-1})C_A$	40 °C
$CH_2OCH_2 \rightarrow CH_4 + CO$	$(-r_A) = 2.2 \times 10^{-2}(\text{min}^{-1})C_A$	450 °C
$RCH_2OH \rightarrow RCHO + H_2$	$(-r_A) = 0.082(\text{lt} gmol^{-1}\text{min}^{-1})C_A^2$	1000 °C

- *Method of total pressure*: This method is similar to the integral method and is used exclusively for gas phase reactions in which there is a variation of the number of moles. The data used are time and total pressure of the system.
- *Method of half-life time*: Unlike the other methods, which use experimental data of time and another property of the system, the method of half-life time requires the value of the time that is needed for the initial concentration of the reactant to reduce to half of its value. To have different experimental data, the reaction has to be conducted with different initial concentrations of the reactant.

These methods are described in greater detail in this chapter, and the resulting final equations that have to be used for mathematical treatment of the kinetic data are highlighted.

2.1 Integral Method

The integral method is easy and simple to apply. It is recommended for relatively simple kinetic expressions corresponding to elemental reactions or when the experimental data are so disperse that reaction rates cannot be calculated with enough accuracy to allow the use of the differential method.

The integral method loses its easiness when the evaluated kinetic equation is not simple to integrate. In addition, due to its iterative nature, for non-elemental reactions the solution may require several iterations.

The integral method always requires the integration of the selected kinetic expression. For irreversible reactions, the following general procedure is followed:

1) Transform the experimental data into molar concentration or conversion of the limiting reactant using the equations described in Chapter 1.

2) Determine the type of reacting system, either constant density or variable density, by taking into account the stoichiometry, type of reaction and operating conditions, and select the reaction rate expression to be used:

Constant Density

$$(-r_A) = -\frac{dC_A}{dt} = kC_A^n \tag{2.3}$$

$$(-r_A) = C_{Ao}\frac{dx_A}{dt} = kC_{Ao}^n(1-x_A)^n \tag{2.4}$$

Variable Density

$$(-r_A) = \frac{C_{Ao}}{1+\varepsilon_A x_A}\frac{dx_A}{dt} = \frac{k\,C_{Ao}^n(1-x_A)^n}{(1+\varepsilon_A x_A)^n} \tag{2.5}$$

3) Assume a reaction order n ($0 \le n \le 3$). It is recommended to start the iterations by considering that the reaction is elemental.
4) Substitute the assumed reaction order in Eqs. (2.3)–(2.5), depending on the case.
5) Separate the variables, and integrate from the initial state ($t = 0$, $C_A = C_{Ao}$, $x_A = 0$) to the final state (t, C_A, x_A).
6) Verify if the assumed reaction order is correct by means of any of the following approaches:
   A) Substitution or analytical method
      - Calculate the reaction rate coefficient k from the integrated equation obtained in point 5.
      - If k exhibits similar values without any tendency, the proposed reaction order is correct, and the reaction rate coefficient is evaluated as the average of all the calculated values.
   B) Graphical method
      - Transform the integrated equation obtained in point 5 into a linear form, in such a way that the x-axis contains the time (t) and the y-axis contains a function of concentration or conversion, $f(C_A)$ or $f(x_A)$.
      - Plot the experimental data and if they fit a straight line, the assumed reaction order is correct, and the reaction rate coefficient is determined from the slope of the straight line.
7) If any of the approaches fail, go to point 3 and assume a new reaction order.

2.1.1 Reactions of Zero Order

Systems at Constant Density
For $n = 0$, Eq. (2.3) is:

$$(-r_A) = -\frac{dC_A}{dt} = k \tag{2.6}$$

Integrating from $t = 0$ and $C_A = C_{Ao}$ to t and C_A:

$$-\int_{C_{Ao}}^{C_A} dC_A = k\int_0^t dt$$

$$C_{Ao} - C_A = kt$$

And, as a function of conversion x_A:

$$C_{Ao}x_A = kt$$

And, finally, the following expression of the straight line is obtained:

$$C_{Ao}x_A = C_{Ao} - C_A = kt \tag{2.7}$$

The reaction rate coefficient k is obtained directly with the slope of Eq. (2.7) by plotting t versus $C_{Ao}x_A$ or $(C_{Ao} - C_A)$, as illustrated in Figure 2.1.

The value of k can also be calculated from Eq. (2.7) as follows:

$$k = \frac{C_{Ao} - C_A}{t} = \frac{C_{Ao}x_A}{t} \tag{2.8}$$

Systems at Variable Density
For $n = 0$, Eq. (2.5) is:

$$(-r_A) = \frac{C_{Ao}}{1 + \varepsilon_A x_A}\frac{dx_A}{dt} = k \tag{2.9}$$

Separating variables and integrating from $t = 0$ and $x_A = 0$ to t and x_A:

$$C_{Ao}\int_0^{x_A} \frac{dx_A}{1 + \varepsilon_A x_A} = k\int_0^t dt$$

$$\frac{C_{Ao}}{\varepsilon_A}\ln(1 + \varepsilon_A x_A) = kt$$

Combining this equation with Eq. (1.56), where $V = V_o(1 + \varepsilon_A x_A)$:

$$\ln(1 + \varepsilon_A x_A) = \ln\left(\frac{V}{V_o}\right) = \left(\frac{k\varepsilon_A}{C_{Ao}}\right)t \tag{2.10}$$

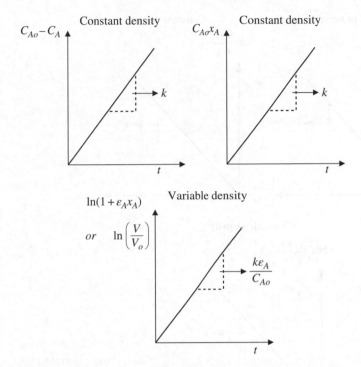

Figure 2.1 Graphical integral method for irreversible reactions of one component of zero order.

By plotting t versus $ln(1 + \varepsilon_A x_A)$ or $ln(V/V_o)$, the value of k can be calculated from the slope of the straight line $(k\varepsilon_A/C_{Ao})$. Figure 2.1 also shows Eq. (2.10) in a graphical form.

The value of k can also be obtained from Eq. (2.10) as follows:

$$k = \frac{C_{Ao}}{\varepsilon_A t} \ln(1 + \varepsilon_A x_A) = \frac{C_{Ao}}{\varepsilon_A t} \ln\left(\frac{V}{V_o}\right) \tag{2.11}$$

2.1.2 Reactions of the First Order

Systems at Constant Density
For $n = 1$, Eq. (2.3) is:

$$(-r_A) = -\frac{dC_A}{dt} = kC_A$$

$$-\int_{C_{Ao}}^{C_A} \frac{dC_A}{C_A} = k\int_0^t dt \tag{2.12}$$

$$\ln\frac{C_{Ao}}{C_A} = kt$$

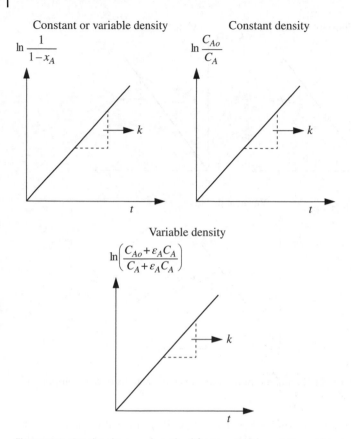

Figure 2.2 Graphical integral method for irreversible reactions of one component of the first order.

Since $C_A = C_{Ao}(1 - x_A)$:

$$\ln\frac{1}{1-x_A} = kt$$

And, finally, the following expression of the straight line is obtained:

$$\ln\frac{C_{Ao}}{C_A} = \ln\frac{1}{1-x_A} = kt \qquad (2.13)$$

This equation is illustrated in Figure 2.2. The value of k can be directly determined from the slope of the straight line. k can also be obtained from Eq. (2.13):

$$k = \frac{1}{t}\ln\frac{C_{Ao}}{C_A} = \frac{1}{t}\ln\frac{1}{1-x_A} \qquad (2.14)$$

Systems at Variable Density
For $n = 1$, Eq. (2.5) is:

$$(-r_A) = \frac{C_{Ao}}{1 + \varepsilon_A x_A} \frac{dx_A}{dt} = k \frac{C_{Ao}(1 - x_A)}{1 + \varepsilon_A x_A} \tag{2.15}$$

Separating variables and integrating:

$$\int_0^{x_A} \frac{dx_A}{1 - x_A} = k \int_0^t dt$$

$$\ln\left(\frac{1}{1 - x_A}\right) = kt \tag{2.16}$$

Since:

$$C_A = \frac{C_{Ao}(1 - x_A)}{1 + \varepsilon_A x_A}$$

By combining these last two equations, the following expression of the straight line is obtained:

$$x_A = \frac{C_{Ao} - C_A}{C_{Ao} + \varepsilon_A C_A}$$

$$\ln\left(\frac{C_{Ao} + \varepsilon_A C_A}{C_A + \varepsilon_A C_A}\right) = kt \tag{2.17}$$

k can be calculated with:

$$k = \frac{1}{t} \ln \frac{1}{1 - x_A} = \frac{1}{t} \ln\left(\frac{C_{Ao} + \varepsilon_A C_A}{C_A + \varepsilon_A C_A}\right) \tag{2.18}$$

As in previous cases, with the slope of Eqs. (2.16) or (2.17), the value of k is determined (Figure 2.2).

It is observed that Eq. (2.16) at variable density as a function of x_A is the same as Eq. (2.13) at variable density. In addition, when k is calculated with any of the two equations, the value of C_{Ao} is not needed, which only occurs for the first order of the reaction.

2.1.3 Reaction of the Second Order

Systems at Constant Density
For $n = 2$, Eq. (2.3) is:

$$(-r_A) = -\frac{dC_A}{dt} = kC_A^2 \tag{2.19}$$

Separating variables and integrating:

$$-\int_{C_{Ao}}^{C_A} \frac{dC_A}{C_A^2} = k\int_0^t dt$$

$$\frac{1}{C_A} - \frac{1}{C_{Ao}} = kt \tag{2.20}$$

And as a function of x_A:

$$\frac{1}{C_{Ao}}\left(\frac{1}{1-x_A}-1\right) = kt$$

$$\frac{x_A}{1-x_A} = kC_{Ao}t \tag{2.21}$$

The slope of Eq. (2.20) is equal to k, and for Eq. (2.21) it is equal to kC_{Ao} (Figure 2.3).

The value of k can also be calculated with:

$$k = \frac{1}{t}\left(\frac{1}{C_A}-\frac{1}{C_{Ao}}\right) = \frac{x_A}{C_{Ao}t(1-x_A)} \tag{2.22}$$

Systems at Variable Density
For $n = 2$, Eq. (2.5) is:

$$(-r_A) = \frac{C_{Ao}}{1+\varepsilon_A x_A}\frac{dx_A}{dt} = \frac{kC_{Ao}^2(1-x_A)^2}{(1+\varepsilon_A x_A)^2} \tag{2.23}$$

Separating variables:

$$\int_0^{x_A} \frac{1+\varepsilon_A x_A}{(1-x_A)^2}dx_A = kC_{Ao}\int_0^t dt = kC_{Ao}t$$

Dividing the integral in the left-hand side in two integrals:

$$\int_0^{x_A} \frac{1+\varepsilon_A x_A}{(1-x_A)^2}dx_A = \int_0^{x_A} \frac{dx_A}{(1-x_A)^2} + \varepsilon_A\int_0^{x_A} \frac{x_A}{(1-x_A)^2}dx_A$$

The integration gives:

$$\frac{x_A}{1-x_A} + \varepsilon_A\left[\frac{x_A}{1-x_A}+\ln(1-x_A)\right] = \frac{(1+\varepsilon_A)x_A}{1-x_A}+\varepsilon_A\ln(1-x_A)$$

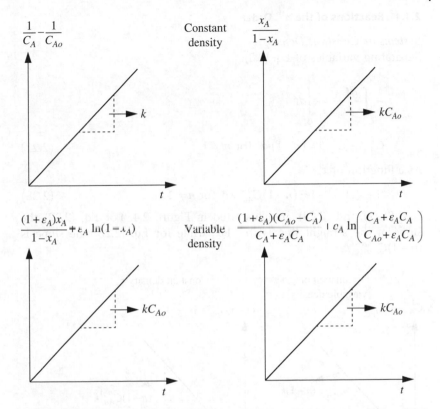

Figure 2.3 Graphical integral method for irreversible reactions of one component of the second order.

And, finally:

$$\frac{(1+\varepsilon_A)x_A}{1-x_A} + \varepsilon_A \ln(1-x_A) = kC_{Ao}t \tag{2.24}$$

which can also be expressed as a function of C_A:

$$\frac{(1+\varepsilon_A)(C_{Ao}-C_A)}{C_A + \varepsilon_A C_A} + \varepsilon_A \ln\left(\frac{C_A + \varepsilon_A C_A}{C_{Ao} + \varepsilon_A C_A}\right) = kC_{Ao}t \tag{2.25}$$

Figure 2.3 also shows Eqs. (2.24) and (2.25). In both cases, the slope of the straight line (kC_{Ao}) gives the value of k, which can also be calculated with:

$$k = \frac{1}{C_{Ao}t}\left[\frac{(1+\varepsilon_A)x_A}{1-x_A} + \varepsilon_A \ln(1-x_A)\right]$$

$$= \frac{1}{C_{Ao}t}\left[\frac{(1+\varepsilon_A)(C_{Ao}-C_A)}{C_A + \varepsilon_A C_A} + \varepsilon_A \ln\left(\frac{C_A + \varepsilon_A C_A}{C_{Ao} + \varepsilon_A C_A}\right)\right] \tag{2.26}$$

2.1.4 Reactions of the nth Order

Systems at Constant Density
Separating variables of Eq. (2.3):

$$-\int_{C_{Ao}}^{C_A} \frac{dC_A}{C_A^n} = k\int_0^t dt$$

$$C_A^{1-n} - C_{Ao}^{1-n} = (n-1)kt \quad \text{for } n \neq 1 \tag{2.27}$$

As a function of x_A:

$$(1-x_A)^{1-n} - 1 = (n-1)C_{Ao}^{n-1}kt \quad \text{for } n \neq 1 \tag{2.28}$$

Eqs. (2.27) and (2.28) are illustrated in Figure 2.4. For Eq. (2.27) the slope of the straight line is $(n-1)k$, while for Eq. (2.28) the slope is $(n-1)C_{Ao}^{n-1}k$.

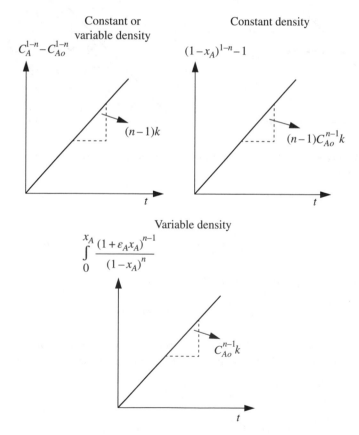

Figure 2.4 Graphical integral method for irreversible reactions of one component of the nth order.

For this case, a reaction order has to be assumed, then it is substituted in these equations, and finally the plot is prepared with experimental data. The value of k is given by:

$$k = \frac{1}{(n-1)t}\left[C_A^{1-n} - C_{Ao}^{1-n}\right] = \frac{1}{(n-1)\,C_{Ao}^{n-1}t}\left[(1-x_A)^{1-n} - 1\right] \quad \text{for } n \neq 1$$

$$(2.29)$$

It is important to mention that the equations for the nth order are valid for $n \neq 1$.

Systems at Variable Density
Separating variables of Eq. (2.5):

$$\int_0^{x_A} \frac{(1+\varepsilon_A x_A)^{n-1}}{(1-x_A)^n}dx_A = C_{Ao}^{n-1}kt \qquad (2.30)$$

This equation is not easy to solve analytically. To do that, it is necessary first to assume a value of n and then integrate the resulting expression. In any case, the final equation is transformed into a linear equation, whereby the value of k is obtained from the slope, as observed in Figure 2.4.

Example 2.1 For the following pyrolysis reaction of dimethylether carried out at 504 °C in a batch reactor operated at constant density, the experimental data reported in Table 2.2 were obtained. Find the rate equation for this reaction.

$$CH_3OCH_3 \rightarrow CH_4 + H_2 + CO \quad (A \rightarrow Products)$$

Table 2.2 Data and results of Example 2.1.

t (sec)	P (KPa)	x_A	$k \times 10^4$ ($n = 1$)	$k \times 10^2$ ($n = 2$)
0	41.6	0	–	–
390	54.4	0.1538	4.283	7.241
777	65.1	0.2825	4.272	7.868
1195	74.9	0.4002	4.278	8.673
3155	103.9	0.7488	4.370	10.674
Average	–	–	4.301	–

Solution

First, it is necessary, in order to transform the data of total pressure into conversion or concentration, to use the equations developed previously in this chapter. This can be done with Eq. (1.42) because the system is at constant density. To apply Eq. (1.42), the value of ε_A is needed, which can be calculated with Eq. (1.40). It is also known that at $t = 0$, the initial total pressure is $P_o = 41.6$ KPa.

$$\varepsilon_A = \frac{y_{Ao}\Delta n}{a} = \frac{(1)(3-1)}{1} = 2$$

For instance, for $t = 390$ sec and $P = 54.4$ KPa:

$$x_A = \frac{P - P_o}{\varepsilon_A P_o} = \frac{P - 41.6}{(2)(41.6)} = \frac{P}{83.2} - 0.5 = \frac{54.4}{83.2} - 0.5 = 0.1538$$

All values of x_A calculated for each value of P with the previous equation are presented in Table 2.2.

To use the integral method, it is necessary to assume a value of reaction order. The first supposition is considering that the reaction is elemental, so that $n = 1$. Eq. (2.14) as a function of x_A is used to calculate k. For instance, for $t = 390$ sec, $x_A = 0.1538$:

$$k = \frac{1}{t}\ln\frac{1}{1-x_A} = \frac{1}{390}\ln\frac{1}{1-0.1538} = 4.283 \times 10^{-4}\,\text{sec}^{-1}$$

The other results are summarized in Table 2.2. The values of k calculated with the previous equation are similar and do not exhibit any tendency, so the assumed order is correct ($n = 1$).

As an additional example for comparison, the values of k were calculated for $n = 2$ with Eq. (2.22). To use this equation, the value of the initial concentration C_{Ao} is needed, which is calculated with Eq. (1.33):

$$C_{Ao} = \frac{p_{Ao}}{RT} = \frac{P_o y_{Ao}}{RT}$$

$$= \frac{(41.6\,KPa)(1)}{\left(0.08205\dfrac{atm\,lt}{gmolK}\right)(504 + 273.15)K}\left(\frac{1\,atm}{101.325\,KPa}\right)$$

$$= 6.439 \times 10^{-3}\frac{gmol}{lt}$$

$$k = \frac{x_A}{C_{Ao}t(1-x_A)} = \frac{0.1538}{(6.439 \times 10^{-3})(390)(1-0.1538)}$$

$$= 7.241 \times 10^{-2}\frac{lt}{gmol\,sec}$$

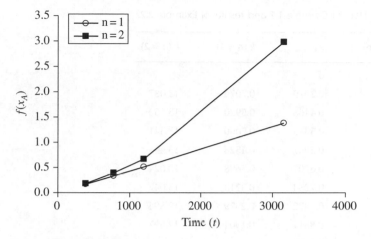

Figure 2.5 Results of Example 2.1.

The results are summarized in Table 2.2. It is seen that the values indeed show a clear tendency to increase as time also increases; therefore, the reaction does not follow the second order.

Figure 2.5 confirms that the reaction follows the first order, since the experimental points fit a straight line with a correlation coefficient of 0.9999. The second order of the reaction does not fit the experimental data.

The slope of the straight line obtained by linear regression gives a value of $k = 4.301 \times 10^{-4}$ sec^{-1}. The value of k can also be obtained as the average value of those calculated with Eq. (2.14), which is $k = 4.404 \times 10^{-4}$ sec^{-1}. As it is observed, both approaches report similar values of k, although the linear regression method is more accurate than the substitution method.

Finally, the kinetic model is:

$$(-r_A) = 4.301 \times 10^{-4} \left(\text{sec}^{-1}\right) C_A$$

Example 2.2   Find the kinetic model of the reaction of Examples 1.4 and 1.7 at variable density.

Solution

The data of x_A versus t calculated in Example 1.7 are reported in Table 2.3. In addition, the following information is known for the reaction:

Stoichiometry of the reaction: $A \rightarrow R + S$
Feed composition: 100% A

Table 2.3 Data of Example 1.7 and results of Example 2.2.

Time (min)	x_A	k ($n = 1$)	k ($n = 2$)
0	0		
0.5	0.2979	0.7074	13.087
1.0	0.4468	0.5920	13.530
1.5	0.5319	0.5060	13.341
2.0	0.5957	0.4528	13.495
3.0	0.6702	0.3698	13.024
4.0	0.7234	0.3213	13.042
6.0	0.7872	0.2579	12.894
10.0	0.8511	0.1904	12.597
∞	1.0000	–	–
Average	–		13.126

$$P = P_{atm} + 1000\,mmHg = 1760\,mmHg$$
$$T = 100°C + 273.15 = 373.15K$$

With this information, the initial concentration of A and ε_A can be calculated:

$$C_{Ao} = \frac{p_{Ao}}{RT} = \frac{P_o y_{Ao}}{RT} = \frac{(1760\,mmHg)(1)}{\left(62.361\dfrac{mmHg\,lt}{gmolK}\right)(373.15\,K)}$$

$$= 7.563 \times 10^{-2}\frac{gmol}{lt}$$

$$\varepsilon_A = \frac{y_{Ao}\Delta n}{a} = \frac{(1)(2-1)}{1} = 1$$

To use the integral method, it is necessary to assume a value of n. Starting with the order of the elemental reaction ($n = 1$), Eq. (2.18) as a function of x_A is used to calculate k. For instance, for the first experimental point:

$$k = \frac{1}{t}\ln\frac{1}{1-x_A} = \frac{1}{0.5}\ln\frac{1}{1-0.2979} = 0.7074\,min^{-1}$$

The other calculations are summarized in Table 2.3. It is observed that the values of k present a descending tendency; thus, the reaction does not follow the first order. The second supposition is $n = 2$, and Eq. (2.26) as a function of x_A is used:

$$k = \frac{1}{C_{Ao}t}\left[\frac{(1+\varepsilon_A)x_A}{1-x_A} + \varepsilon_A \ln(1-x_A)\right]$$

$$= \frac{1}{(7.563 \times 10^{-2})(0.5)}\left[\frac{(1+1)(0.2979)}{1-0.2979} + (1)\ln(1-0.2979)\right]$$

$$= 13.087 \frac{lt}{gmol\ min}$$

The other calculations are reported in Table 2.3. In this case, the second-order values of k do not show any tendency, and they are more or less constant, so that the kinetic model is:

$$(-r_A) = 13.126 \left(gmol^{-1}lt\ min^{-1}\right)C_A^2$$

2.2 Differential Method

Unlike the integral method, in the differential method the kinetic expression is not integrated, but it is used in its differential form (Eq. 2.3) and transformed into a linear equation by means of logarithms.

The differential method is useful for more complicated cases, but it requires a greater number of experimental data to achieve more accurate results regarding the reaction rate coefficient and reactor order. The main advantage of the differential method over the integral method is that the former directly gives the values of n and k.

Starting with Eqs. (2.3)–(2.5), the following general expressions can be derived to apply the differential method.

Constant Density
Equations as functions of concentration:

$$(-r_A) = -\frac{dC_A}{dt} = kC_A^n$$

$$\ln\left(-\frac{dC_A}{dt}\right) = \ln k + n \ln C_A$$

$$(2.31)$$

Equations as functions of conversion:

$$(-r_A) = C_{Ao}\frac{dx_A}{dt} = kC_{Ao}^n(1-x_A)^n$$

$$\frac{dx_A}{dt} = kC_{Ao}^{n-1}(1-x_A)^n$$

$$\ln\left(\frac{dx_A}{dt}\right) = \ln\left(kC_{Ao}^{n-1}\right) + n\ln(1-x_A)$$

$$(2.32)$$

Variable Density

Equations as functions of concentration:

$$(-r_A) = -\frac{dC_A}{dt} = kC_A^n$$

$$\ln\left(-\frac{dC_A}{dt}\right) = \ln k + n\ln C_A \tag{2.33}$$

Equations as functions of conversion:

$$(-r_A) = \frac{C_{Ao}}{1+\varepsilon_A x_A}\frac{dx_A}{dt} = \frac{k\,C_{Ao}^n(1-x_A)^n}{(1+\varepsilon_A x_A)^n}$$

$$\frac{dx_A}{dt} = \frac{k\,C_{Ao}^{n-1}(1-x_A)^n}{(1+\varepsilon_A x_A)^{n-1}}$$

$$\ln\left(\frac{dx_A}{dt}\right) = \ln\left[\frac{k\,C_{Ao}^{n-1}}{(1+\varepsilon_A x_A)^{n-1}}\right] + n\ln(1-x_A) \tag{2.34}$$

Eqs. (2.31)–(2.34) are straight lines of the type $y = b + mx$. For instance, Figure 2.6 shows the graphical representation of Eq. (2.31). For the four cases, the slope of the straight lines gives the value of n.

At constant density, k can be determined from the intercept of the straight line: $\ln k$ for the equation as a function of concentration (Eq. 2.31), and $\ln (kC_{Ao}^{n-1})$ for the equation as a function of conversion (Eq. 2.32).

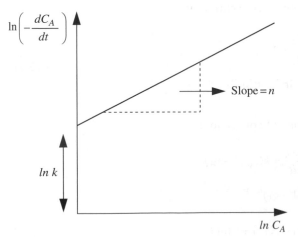

Figure 2.6 Differential method for analysis of kinetic data with Eq. (2.31).

At variable density, k is also calculated from the intercept of the straight line in the case of the equation as a function of concentration (Eq. 2.33). However, for the equation as a function of conversion (Eq. 2.34), it is not possible to calculate k from the intercept of the straight line. One way to determine k is to use the integral method once the value of n has been determined with the differential method. Another way is to transform the data of conversion into concentration with Eq. (1.58), and use Eq. (2.33) to calculate k as explained before.

The application of these equations requires knowledge of the data of $ln(-dC_A/dt)$ or $ln(dx_A/dt)$ and lnC_A or lnx_A.

The evaluation of the derivatives can be done by means of differentiation of the data of concentration C_A or conversion x_A and time t.

There are different approaches to perform this differentiation, either numerical or graphical; the most used ones are described in the remainder of this section. For each case, the equations are derived for C_A, but they are also valid for x_A or any other property used for calculating the reaction rate.

2.2.1 Numerical Differentiation

2.2.1.1 Method of Approaching the Derivatives $(-dC_A/dt)$ to $(\Delta C_A/\Delta t)$ or (dx_A/dt) to $(\Delta x_A/\Delta t)$

This method considers that the derivatives $(-dC_A/dt)$ or (dx_A/dt) can be represented as the variations $(\Delta C_A/\Delta t)$ or $(\Delta x_A/\Delta t)$; however, to obtain accurate results, sufficient experimental data of t and C_A or x_A are needed.

This consideration simplifies the use of the differential methods, and Eqs. (2.31)–(2.34) are transformed into:

Constant Density
Equations as functions of concentration:

$$\ln \frac{\Delta C_A}{\Delta t} = \ln k + n \ln C_{Ap} \tag{2.35}$$

Equations as functions of conversion:

$$\ln \frac{\Delta x_A}{\Delta t} = \ln \left(k C_{Ao}^{n-1} \right) + n \ln \left(1 - x_{Ap} \right) \tag{2.36}$$

Variable Density
Equations as functions of concentration:

$$\ln \frac{\Delta C_A}{\Delta t} = \ln k + n \ln C_{Ap} \tag{2.37}$$

Table 2.4 Procedure to calculate $\Delta C_A/\Delta t$.

Time	C_A	ΔC_A	Δt	$\Delta C_A/\Delta t$	C_{Ap}
$t_0 = 0$	C_{A0}				
		$C_{A0} - C_{A1}$	$t_1 - t_0$	$\dfrac{C_{Ao} - C_{A1}}{t_1 - t_o}$	$\dfrac{C_{Ao} + C_{A1}}{2}$
t_1	C_{A1}				
		$C_{A1} - C_{A2}$	$t_2 - t_1$	$\dfrac{C_{A1} - C_{A2}}{t_2 - t_1}$	$\dfrac{C_{A1} + C_{A2}}{2}$
t_2	C_{A2}				
		$C_{A2} - C_{A3}$	$t_3 - t_2$	$\dfrac{C_{A2} - C_{A3}}{t_3 - t_2}$	$\dfrac{C_{A2} + C_{A3}}{2}$
t_3	C_{A3}				
		$C_{A3} - C_{A4}$	$t_4 - t_3$	$\dfrac{C_{A3} - C_{A4}}{t_4 - t_3}$	$\dfrac{C_{A3} + C_{A4}}{2}$
t_4	C_{A4}				
t_{n-1}	C_{An-1}				
		$C_{An-1} - C_{An}$	$t_n - t_{n-1}$	$\dfrac{C_{An-1} - C_{An}}{t_n - t_{n-1}}$	$\dfrac{C_{An-1} + C_{An}}{2}$
t_n	C_{An}				

Equations as functions of conversion:

$$\ln\frac{\Delta x_A}{\Delta t} = \ln\left[\frac{k\, C_{Ao}^{n-1}}{(1 + \varepsilon_A x_A)^{n-1}}\right] + n\ln(1 - x_{Ap}) \tag{2.38}$$

where C_{Ap} and x_{Ap} represent the average values of the two values of C_A and x_A that were used to calculate ΔC_A and Δx_A.

Table 2.4 illustrates the procedure with data of C_A to determine the parameters that are needed in Eqs. (2.35)–(2.38). Once the values of $\Delta C_A/\Delta t$ and C_{Ap} are known, and using Eq. (2.35), the values of n and k can be determined graphically as illustrated in Figure 2.7.

2.2.1.2 Method of Finite Differences

The method of finite differences can use the following formulae of three-point differentiation to evaluate the derivative $(-dC_A/dt)$.

For the first datum:

$$\left(\frac{dC_A}{dt}\right)_{i=0} = \left(\frac{1}{2\Delta t}\right)(-3C_{Ao} + 4C_{A1} - C_{A2}) \tag{2.39}$$

Figure 2.7 Method of numerical differentiation by approaching the derivative $(-dC_A/dt)$ to $(\Delta C_A/\Delta t)$.

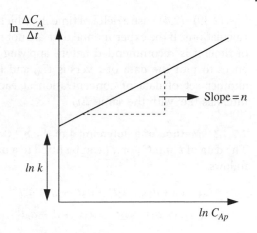

For intermediate data:

$$\left(\frac{dC_A}{dt}\right)_{0<i<n} = \left(\frac{1}{2\Delta t}\right)(C_{Ai+1} - C_{Ai-1}) \qquad (2.40)$$

For the last datum:

$$\left(\frac{dC_A}{dt}\right)_{i=n} = \left(\frac{1}{2\Delta t}\right)(C_{An-2} - 4C_{An-1} + 3C_{An}) \qquad (2.41)$$

Table 2.5 presents an example of the formulae that are used to apply the method of finite differences for the case of five experimental points, including the initial datum at zero time.

Table 2.5 Method of finite differences for five experimental points.

i	t	C_A	$-dC_A/dt$
0	t_1	C_{A1}	$\left(\dfrac{dC_A}{dt}\right)_0 = \left(\dfrac{1}{2\Delta t}\right)(-3C_{Ao} + 4C_{A1} - C_{A2})$
1	t_2	C_{A2}	$\left(\dfrac{dC_A}{dt}\right)_1 = \left(\dfrac{1}{2\Delta t}\right)(C_{A2} - C_{Ao})$
2	t_3	C_{A3}	$\left(\dfrac{dC_A}{dt}\right)_2 = \left(\dfrac{1}{2\Delta t}\right)(C_{A3} - C_{A1})$
3	t_4	C_{A4}	$\left(\dfrac{dC_A}{dt}\right)_3 = \left(\dfrac{1}{2\Delta t}\right)(C_{A4} - C_{A2})$
4	t_5	C_{A5}	$\left(\dfrac{dC_A}{dt}\right)_4 = \left(\dfrac{1}{2\Delta t}\right)(C_{A2} - 4C_{A3} + 3C_{A4})$

Eqs. (2.39)–(2.41) use a delta of time Δt, which must be the same for all the calculations. If the experimental data are not measured at regular intervals of time, it is recommended before applying the method of finite differences to plot the data of t versus C_A, and from this graph to generate another set of data of concentration at each time considering regular intervals (i.e. with the same Δt).

2.2.1.3 Method of a Polynomial of the nth Order

The data of t and C_A or x_A can be fitted to a polynomial of the nth order as follows:

$$C_A = a_0 + a_1 t + a_2 t^2 + a_3 t^3 + \dots + a_n t^n \tag{2.42}$$

$$x_A = a_0 + a_1 t + a_2 t^2 + a_3 t^3 + \dots + a_n t^n \tag{2.43}$$

The constants a_i are obtained by polynomial regression analysis.

The order of the polynomial is fixed based on the correlation coefficient r between the experimental data (t vs. C_A or x_A) and those calculated with the polynomial. The order of the polynomial that exhibits the higher value of r would be the more suitable to be used in Eqs. (2.42) or (2.43). It should be remembered that the order of the polynomial must not exceed the number of data.

Once the experimental data are fitted to a polynomial, the derivative of Eqs. (2.42) and (2.43) with respect to time are used to obtain (dC_A/dt) or $(-dx_A/dt)$:

$$\frac{dC_A}{dt} = a_1 + 2a_2 t + 3a_3 t^2 + \dots + na_n t^{n-1} \tag{2.44}$$

$$\frac{dx_A}{dt} = a_1 + 2a_2 t + 3a_3 t^2 + \dots + na_n t^{n-1} \tag{2.45}$$

And, finally, for all experimental data the corresponding values of the derivatives (dC_A/dt) or (dx_A/dt) are calculated with Eqs. (2.44) or (2.45).

This method is fast and easy to implement once the polynomial has been obtained.

2.2.2 Graphical Differentiation

2.2.2.1 Method of Area Compensation

The method of graphical differentiation by area compensation is an extension of the method of numerical differentiation by approaching the derivatives $(-dC_A/dt)$ to $(\Delta C_A/\Delta t)$ or (dx_A/dt) to $(\Delta x_A/\Delta t)$, because the values of $(\Delta C_A/\Delta t)$ or $(\Delta x_A/\Delta t)$ obtained numerically (Table 2.4) have to be plotted with respect to t in the form of rectangles.

For instance, from Table 2.4, $(\Delta C_A/\Delta t)_1$ was calculated with the data of t_1 and t_2. Then, when plotting t versus $(\Delta C_A/\Delta t)$, the area of the formed rectangle must have as a base the difference between t_1 and t_2.

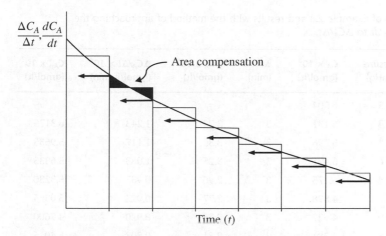

Figure 2.8 Method of graphic differentiation by area compensation.

The next step is to draw an average curve over the area of rectangles obtained, trying as far as possible to ensure that the area being left out is similar to the area that does not belong to the rectangle, but that it is indeed being considered within the area under the curve (i.e. the areas are compensated for).

Finally, the values of (dC_A/dt) are taken from the plotted curve for each value of t, as illustrated in Figure 2.8.

Example 2.3 In the decomposition of di-tert-butyl peroxide to ketone and ethane:

$$(CH_3)_3COOC(CH_3)_3 \rightarrow 2(CH_3)_2CO + C_2H_6 \ (A \rightarrow 2R + S)$$

The experimental data reported in Table 2.6 were obtained at 154.6 °C. Find the reaction order and the reaction rate coefficient.

Solution

To convert the data of total pressure (P) to concentration (C_A), Eq. (1.52) is used. Before, the values of ε_A and C_{Ao} are needed:

$$T = 154.6 + 273.15 = 427.75 K$$

$$\varepsilon_A = \frac{y_{Ao}\Delta n}{a} = \frac{(1)(3-1)}{1} = 2$$

$$C_{Ao} = \frac{p_{Ao}}{RT} = \frac{P_o y_{Ao}}{RT} = \frac{(173.5 \, mmHg)(1)}{\left(62.36 \dfrac{mmHg \, lt}{gmolK}\right)(427.75)K}$$

$$= 6.504 \times 10^{-3}\frac{gmol}{lt}$$

Table 2.6 Data of Example 2.3 and results with the method of approaching the derivative $-dC_A/dt$ to $\Delta C_A/\Delta t$.

Time (min)	Pressure (mmHg)	$C_A \times 10^3$ (gmol/lt)	Δt (min)	$\Delta C_A \times 10^4$ (gmol/lt)	$\Delta C_A/\Delta t \times 10^4$ (gmol/lt min)	$C_{Ap} \times 10^3$ (gmol/lt)
0	173.5	6.504				
3	193.4	6.131	3	3.73	1.243	6.3175
6	211.3	5.796	3	3.35	1.117	5.9635
9	228.6	5.471	3	3.25	1.083	5.6335
12	244.4	5.175	3	2.96	0.987	5.3230
15	259.2	4.898	3	2.77	0.923	5.0365
18	273.9	4.622	3	2.76	0.920	4.7600
21	286.8	4.381	3	2.31	0.803	5.4015

Once ε_A and C_{Ao} are known, C_A is calculated with Eq. (1.52). For instance, for the first data:

$$C_A = \frac{y_{Ao}}{RT\varepsilon_A}\left[P_o(1+\varepsilon_A) - P\right] = \frac{1}{(62.361)(427.75)(2)}\left[173.5(1+2) - P\right]$$

$$= \frac{1}{53349.84}(520.5 - 193.4) = 6.131 \times 10^{-3}\frac{gmol}{lt}$$

The other results are presented in Table 2.6.

2.2.2.2 Method of Approaching the Derivative ($-dC_A/dt$) to ($\Delta C_A/\Delta t$)

ΔC_A and Δt are calculated as:

$$\Delta C_{A1} = C_{Ao} - C_{A1} = (6.504 \times 10^{-3} - 6.131 \times 10^{-3})$$
$$= 3.73 \times 10^{-4} gmol/lt$$
$$\Delta t_1 = t_1 - t_o = (3-0) = 3\,min$$

$\Delta C_A/\Delta t$ and C_{Ap} are determined with:

$$\left(\frac{\Delta C_A}{\Delta t}\right)_1 = \frac{C_{A1} - C_{A2}}{t_2 - t_1} = \frac{3.73 \times 10^{-4}}{3} = 1.243 \times 10^{-4}\frac{gmol}{lt\,min}$$

$$C_{Ap_1} = \frac{C_{A1} + C_{A2}}{2} = \frac{6.504 \times 10^{-3} + 6.131 \times 10^{-3}}{2} = 6.318 \times 10^{-3}\frac{gmol}{lt}$$

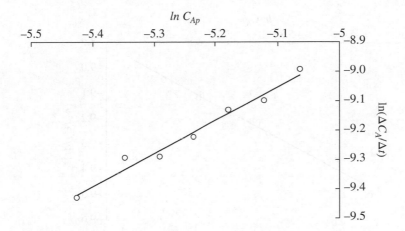

Figure 2.9 Application of the method of approaching the derivative $(-dC_A/dt)$ to $(\Delta C_A/\Delta t)$ for Example 2.3.

The complete results are summarized in Table 2.6. Figure 2.9 shows the application of this method using the linear equation (2.35) in logarithm coordinates with $y = ln(\Delta C_A/\Delta t)$ and $x = lnC_{Ap}$. By means of linear regression analysis, the following results are obtained:

Slope $\qquad = n \qquad = 1.174$

Intercept $\qquad = ln\,k \quad = -3.0606$

Correlation coefficient $\; = r \qquad = 0.9838$

Therefore:

$n = 1.174$

$k = EXP(-3.0606) = 4.686 \times 10^{-2}$

2.2.2.3 Method of Finite Differences

$$\left(\frac{dC_A}{dt}\right)_0 = \left(\frac{1}{2\Delta t}\right)(-3C_{Ao} + 4C_{A1} - C_{A2}) = 1.307 \times 10^{-4}$$

$$\left(\frac{dC_A}{dt}\right)_1 = \left(\frac{1}{2\Delta t}\right)(C_{A2} - C_{Ao}) = 1.18 \times 10^{-4}$$

$$\left(\frac{dC_A}{dt}\right)_2 = \left(\frac{1}{2\Delta t}\right)(C_{A3} - C_{A1}) = 1.10 \times 10^{-4}$$

$$\left(\frac{dC_A}{dt}\right)_3 = \left(\frac{1}{2\Delta t}\right)(C_{A4} - C_{A2}) = 1.035 \times 10^{-4}$$

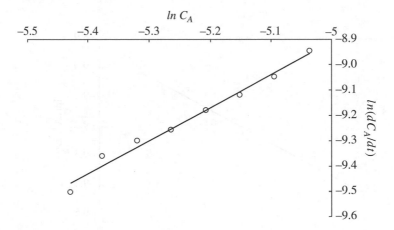

Figure 2.10 Method of finite differences.

$$\left(\frac{dC_A}{dt}\right)_4 = \left(\frac{1}{2\Delta t}\right)(C_{A5} - C_{A3}) = 0.955 \times 10^{-4}$$

$$\left(\frac{dC_A}{dt}\right)_5 = \left(\frac{1}{2\Delta t}\right)(C_{A6} - C_{A4}) = 0.917 \times 10^{-4}$$

$$\left(\frac{dC_A}{dt}\right)_6 = \left(\frac{1}{2\Delta t}\right)(C_{A7} - C_{A5}) = 0.862 \times 10^{-4}$$

$$\left(\frac{dC_A}{dt}\right)_7 = \left(\frac{1}{2\Delta t}\right)(C_{A5} - 4C_{A6} + 3C_{A7}) = 0.745 \times 10^{-4}$$

Applying the linear equation (2.35) with $y = ln(dC_A/dt)$ and $x = ln\, C_A$, the following results are obtained (Figure 2.10):

Slope $= n$ $= 1.293$

Intercept $= ln\, k$ $= -2.4454$

Correlation coefficient $= r$ $= 0.9909$

Therefore:

$n = 1.293$

$k = EXP(-2.4454) = 8.669 \times 10^{-2}$

2.2.2.4 Method of a Polynomial of the nth Order

From the polynomial regression analysis of the data of t versus C_A reported in Table 2.7, it is observed that the 5th-order polynomial exhibits

Table 2.7 Results with the method of polynomium of nth order for Example 2.3.

Order of polynomium	Correlation coefficient	% of error[a]
1	0.99766	0.7982
2	0.99987	0.1630
3	0.99993	0.1330
4	0.99996	0.1034
5	1.00000	0.0354

a) Error = (Experimental value – calculated value) / Experimental value × 100.

the best correlation of the data; however, 3rd-order and 4th-order polynomials also fit the experimental data in an appropriate manner with a correlation coefficient close to unity. To simplify the calculations and employ a smaller number of coefficients, a 3rd-order polynomial will be used. Thus, the values of coefficients of Eq. (2.42) are:

$$a_o = 6.506 \times 10^{-3}$$

$$a_1 = -1.312 \times 10^{-4}$$

$$a_2 = 2.116 \times 10^{-6}$$

$$a_3 = -3.199 \times 10^{-8}$$

Therefore, the final polynomial is:

$$C_A = 6.506 \times 10^{-3} - 1.312 \times 10^{-4}t + 2.116 \times 10^{-6}t^2 - 3.199 \times 10^{-8}t^3$$

And, using Eq. (2.44) for dC_A/dt:

$$\frac{dC_A}{dt} = -1.312 \times 10^{-4} + 4.232 \times 10^{-6}t + -9.597 \times 10^{-8}t^2$$

For each time, the values of dC_A/dt are calculated with the previous equation and are reported in Table 2.8. The calculated (C_A^{calc}) and experimental (C_A^{exp}) values of concentration are also included in Table 2.8 for comparison. A good agreement between both values is observed.

Figure 2.11 shows the fit of the method of a polynomial of the nth order. The final results are:

Slope $= n = 1.135$

Intercept $= ln\,k = -3.2599$

Correlation coefficient $= r = 0.9813$

Table 2.8 Calculation of dC_A/dt with a third-order polynomial for Example 2.3.

Time (min)	$C_A^{exp.} \times 10^3$ (gmol/lt)	$C_A^{calc.} \times 10^3$ (gmol/lt)	$dC_A/dt \times 10^{-4}$ (gmol/lt min)
0	6.504	6.506	1.312
3	6.131	6.131	1.194
6	5.796	5.788	1.093
9	5.471	5.473	1.009
12	5.175	5.181	0.942
15	4.898	4.906	0.893
18	4.622	4.643	0.861
21	4.381	4.388	0.847

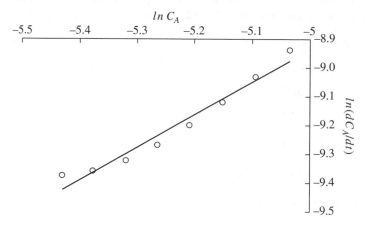

Figure 2.11 Application of the method of a polynomial of the nth order for Example 2.3.

Therefore:

$$n = 1.135$$
$$k = EXP(-3.2599) = 3.839 \times 10^{-2}$$

2.2.2.5 Method of Area Compensation

To apply this graphical method, the results of the method of approaching the derivative dC_A/dt to $\Delta C_A/\Delta t$ are used to prepare a plot of t versus $\Delta C_A/\Delta t$ (Figure 2.12). An average curve is drawn for compensating the

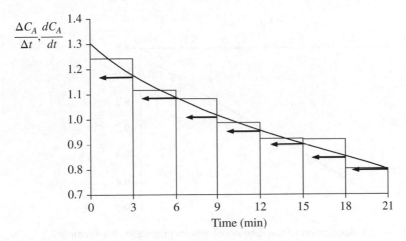

Figure 2.12 Calculation of $\Delta C_A/\Delta t$ at regular intervals of time.

Table 2.9 Calculation of dC_A/dt with the method of area compensation for Example 2.3.

Time (min)	$C_A \times 10^3$ (gmol/lt)	$dC_A/dt \times 10^{-4}$ (gmol/lt min)
0	6.504	1.31
3	6.131	1.19
6	5.796	1.09
9	5.471	1.01
12	5.175	0.95
15	4.898	0.90
18	4.622	0.86
21	4.381	0.81

areas that are inside and outside of such a curve. For each time at regular intervals, the values of $\Delta C_A/\Delta t$ are obtained, which are considered as dC_A/dt (Table 2.9).

Finally, the application of Eq. (2.35) with $y = ln(dC_A/dt)$ and $x = lnC_A$ gives the following results (Figure 2.13):

Slope $\qquad = n \qquad = 1.189$

Intercept $\qquad = ln\,k \quad = -2.984$

Correlation coefficient $\;= r \qquad = 0.9926$

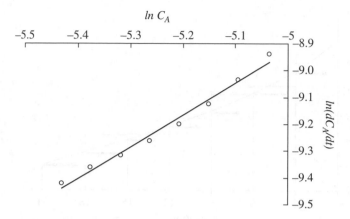

Figure 2.13 Application of the method of area compensation for Example 2.3.

Therefore:

$$n = 1.189$$
$$k = EXP(-2.984) = 5.059 \times 10^{-2}$$

2.2.2.6 Summary of Results

The final results of each of the differentiation methods are presented in Table 2.10. It is observed that the four methods present a correlation coefficient higher than 0.98, and the method of area compensation exhibits the highest value ($r = 0.9926$). The values of n range from 1.135 to 1.293. It is more common that the reaction order is an entire value, so that it can be considered that $n = 1$, and with the integral method (Eq. 2.14) the value of the reaction rate coefficient can be calculated, which results in being $k = 1.871 \times 10^{-2}$. The application of the integral method for this case

Table 2.10 Summary of results of Example 2.3.

Method	Reaction order (n)	Reaction rate coefficient ($k \times 10^2$)	Correlation coefficient (r)
Approaching of the derivative	1.174	4.686	0.9838
Finite differences	1.293	8.669	0.9909
Polynomial of nth order	1.135	3.839	0.9813
Area compensation	1.189	5.059	0.9926

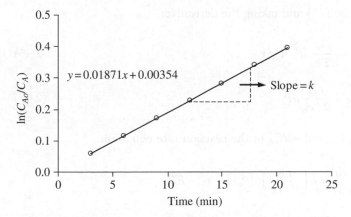

Figure 2.14 Application of the integral method for $n = 1$ for Example 2.3.

is with the following expression, which is represented graphically in Figure 2.14.

$$\ln\frac{C_{Ao}}{C_A} = \ln\frac{1}{1-x_A} = kt$$

Therefore, the kinetic model is:

$$(-r_A) = 1.871 \times 10^{-2}\left(min^{-1}\right)C_A$$

2.3 Method of Total Pressure

In this method, the data of time and total pressure are used to determine the kinetic model. To have variable pressure data, the system has to operate at constant density and temperature, with a gas phase reaction in which there is a change in the number of moles ($\Delta n \neq 0$).

The procedure is similar to that of the integral method, but it varies the integrated equations. Starting with the reaction rate equation for irreversible reactions of one component at constant density (Eq. 2.1):

$$-\frac{dC_A}{dt} = kC_A^n$$

It is necessary to change C_A by P, with Eq. (1.52):

$$C_A = \frac{y_{Ao}}{RT\varepsilon_A}\left[P_o(1+\varepsilon_A) - P\right]$$

Multiplying by (−1) and taking the derivative:

$$-C_A = \frac{y_{Ao}}{RT\varepsilon_A}[P - P_o(1 + \varepsilon_A)] \tag{2.46}$$

$$-dC_A = \frac{y_{Ao}}{RT\varepsilon_A}dP \tag{2.47}$$

Substituting C_A and $-dC_A$ in the reaction rate equation:

$$\frac{y_{Ao}}{RT\varepsilon_A}\frac{dP}{dt} = k\left(\frac{y_{Ao}}{RT\varepsilon_A}\right)^n[P_o(1 + \varepsilon_A) - P]^n$$

$$\frac{dP}{dt} = k\left(\frac{RT\varepsilon_A}{y_{Ao}}\right)^{1-n}[P_o(1 + \varepsilon_A) - P]^n$$

Since:

$$\varepsilon_A = \frac{y_{Ao}\Delta n}{a} \quad \text{or} \quad \frac{\varepsilon_A}{y_{Ao}} = \frac{\Delta n}{a}$$

$$\frac{dP}{dt} = k\left(\frac{RT\Delta n}{a}\right)^{1-n}[P_o(1 + \varepsilon_A) - P]^n \tag{2.48}$$

with:

$$k^* = k\left(\frac{RT\Delta n}{a}\right)^{1-n} \tag{2.49}$$

the general reaction rate equation as a function of total pressure is:

$$\frac{dP}{dt} = k^*[P_o(1 + \varepsilon_A) - P]^n \tag{2.50}$$

where k^* is not the reaction rate coefficient, but just a constant by means of which the value of k can be calculated with Eq. (2.49).

2.3.1 Reactions of Zero Order

For $n = 0$, Eq. (2.50) is:

$$\frac{dP}{dt} = k^*$$

Separating variables and integrating from $t = 0$ and $P = P_o$ to t and P:

$$\int_{P_o}^{P} dP = k^* \int_{0}^{t} dt = k^*t$$

$$P - P_o = k^*t \tag{2.51}$$

And the constant k^* is evaluated as:

$$k^* = \frac{P - P_o}{t} \tag{2.52}$$

2.3.2 Reactions of the First Order

For $n = 1$, Eq. (2.50) is:

$$\frac{dP}{dt} = k^*[P_o(1 + \varepsilon_A) - P]$$

Separating variables and integrating from $t = 0$ and $P = P_o$ to t and P:

$$\int_{P_o}^{P} \frac{dP}{P_o(1 + \varepsilon_A) - P} = k^* \int_{0}^{t} dt = k^*t$$

$$-\ln[P_o(1 + \varepsilon_A) - P]\big|_{P_o}^{P} = k^*t$$

$$-\ln[P_o(1 + \varepsilon_A) - P] + \ln(P_o\varepsilon_A) = k^*t$$

$$\ln\left[\frac{P_o\varepsilon_A}{P_o(1 + \varepsilon_A) - P}\right] = k^*t \tag{2.53}$$

And the constant k^* is evaluated as:

$$k^* = \frac{1}{t}\ln\left[\frac{P_o\varepsilon_A}{P_o(1 + \varepsilon_A) - P}\right] \tag{2.54}$$

2.3.3 Reactions of the Second Order

For $n = 2$, Eq. (2.50) is:

$$\frac{dP}{dt} = k^*[P_o(1 + \varepsilon_A) - P]^2$$

Separating variables and integrating from $t = 0$ and $P = P_o$ to t and P:

$$\int_{P_o}^{P} \frac{dP}{[P_o(1 + \varepsilon_A) - P]^2} = k^* \int_{0}^{t} dt = k^* t$$

$$\frac{1}{P_o(1 + \varepsilon_A) - P} \bigg|_{P_o}^{P} = k^* t$$

$$\frac{1}{P_o(1 + \varepsilon_A) - P} - \frac{1}{P_o \varepsilon_A} = k^* t \tag{2.55}$$

And the constant k^* is evaluated as:

$$k^* = \frac{1}{t} \left[\frac{1}{P_o(1 + \varepsilon_A) - P} - \frac{1}{P_o \varepsilon_A} \right] \tag{2.56}$$

2.3.4 Reactions of the nth Order

Separating variables in Eq. (2.50) and integrating from $t = 0$ and $P = P_o$ to t and P:

$$\int_{P_o}^{P} \frac{dP}{[P_o(1 + \varepsilon_A) - P]^n} = k^* \int_{0}^{t} dt = k^* t$$

$$\frac{1}{n-1} [P_o(1 + \varepsilon_A) - P]^{1-n} \bigg|_{P_o}^{P} = k^* t$$

$$[P_o(1 + \varepsilon_A) - P]^{1-n} - (P_o \varepsilon_A)^{1-n} = (n-1) k^* t \tag{2.57}$$

And the constant k^* is evaluated as:

$$k^* = \frac{1}{(n-1)t} \left\{ [P_o(1 + \varepsilon_A) - P]^{1-n} - (P_o \varepsilon_A)^{1-n} \right\} \tag{2.58}$$

To use Eqs. (2.57) and (2.58), the reaction order needs to be assumed, then substitute it in these equations. Finally, either determine k^* from the slope of the straight line obtained by plotting t versus the left-hand side of Eq. (2.57), or calculate the values of k^* with Eq. (2.58) and check if they are similar without tendency (if they are, calculate the average value, which will be the final value of k^*).

Eqs. (2.51), (2.53), (2.55) and (2.57) are represented in Figure 2.15.

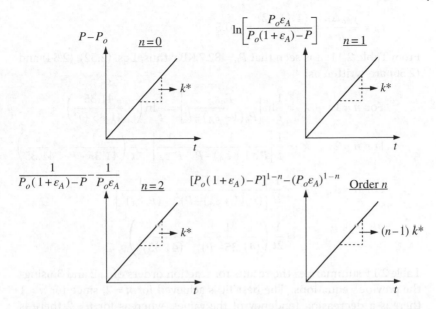

Figure 2.15 Method of total pressure for irreversible reactions of one component.

Example 2.4 The dimerization of trifluorchlorideethylene in gas phase $(2A \rightarrow R)$ was studied by changes in the total pressure. The experimental data obtained at 440 °C are presented in Table 2.11. Find the kinetic model.

Solution

The value of ε_A is required. It is assumed that only the reactant A is in the feed, so that $y_{A0} = 1$.

Table 2.11 Data and results of Example 2.4.

Time (sec)	Pressure (KPa)	$k^* \times 10^3$ $(n = 1)$	$k^* \times 10^5$ $(n = 2)$	$k^* \times 10^6$ $(n = 3)$
0	82.7			
100	71.1	3.292	−9.429	2.725
200	64.0	3.009	−9.983	3.411
300	60.4	2.583	−9.437	3.618
400	56.7	2.477	−10.241	4.574
500	54.8	2.246	−10.033	4.943
Average			-9.825×10^{-5}	

$$\varepsilon_A = \frac{y_{Ao}\Delta n}{a} = \frac{(1)(1-2)}{2} = -0.5$$

From Table 2.11, it is seen that $P_0 = 82.7$ KPa; thus, Eqs. (2.52), (2.54) and (2.56) are written as:

For $n = 1$: $\quad k^* = \frac{1}{t}\ln\left[\frac{P_o\varepsilon_A}{P_o(1+\varepsilon_A)-P}\right] = \frac{1}{t}\ln\left(-\frac{41.35}{41.35-P}\right)$

For $n = 2$: $\quad k^* = \frac{1}{t}\left[\frac{1}{P_o(1+\varepsilon_A)-P} - \frac{1}{P_o\varepsilon_A}\right] = \frac{1}{t}\left(\frac{1}{41.35-P} + \frac{1}{41.35}\right)$

For $n = 3$: $\quad k^* = \frac{1}{2t}\left[\frac{1}{(P_o(1+\varepsilon_A)-P)^2} - \frac{1}{(P_o\varepsilon_A)^2}\right]$

$$= \frac{1}{2t}\left(\frac{1}{(41.35-P)^2} - \frac{1}{(41.35)^2}\right)$$

Table 2.11 summarizes the results for reaction orders of 1, 2 and 3 using the previous equations. The best fit is achieved for $n = 2$, since for $n = 1$ there is a decreasing tendency of the values, whereas for $n = 2$ there is an increasing tendency, so that the average value of k^* is -9.825×10^{-5} sec^{-1} KPa^{-1}.

From Eq. (2.49), the reaction rate coefficient is (note that $R = 8.31$ KPa lt/gmol K):

$$k = \frac{k^*}{\left(\dfrac{a}{RT\Delta n}\right)^{1-n}} = \frac{k^*}{\left(\dfrac{a}{RT\Delta n}\right)^{-1}} = k^*\left(\frac{RT\Delta n}{a}\right)$$

$$= \left(-9.825 \times 10^{-5}\frac{1}{KPa\ sec}\right)\left(\frac{8.31\dfrac{KPa\ lt}{gmolK}(440+273.15)K(1-2)}{2}\right)$$

$$k = 0.2911\frac{lt}{gmol\ sec}$$

And the kinetic model is:

$$(-r_A) = 0.2911(lt/gmol\ sec)C_A{}^2$$

2.3.5 Differential Method with Data of Total Pressure

The values of total pressure can also be used with the differential method to determine the values of the reaction order and reaction rate coefficient

Figure 2.16 Differential method with data of total pressure.

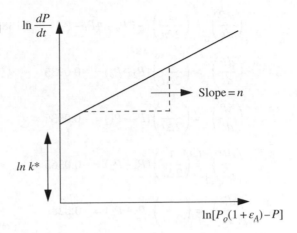

in a direct manner. To do this, Eq. (2.50) is transformed in a linear form by means of logarithms:

$$\ln\frac{dP}{dt} = \ln k^* + n\ln[P_o(1 + \varepsilon_A) - P] \quad \text{For } \varepsilon_A > 0 \tag{2.59}$$

Eq. (2.59) is shown in Figure 2.16. The slope of the straight line directly gives the value of n and the intercept the value of k^*, which is further used with Eq. (2.49) to calculate the value of k. The derivative dP/dt is evaluated with any of the methods described in Sections 2.2.1 and 2.2.2.

For a reaction in which the total pressure diminishes as the reaction proceeds ($P_o > P$), in some cases the term $[P_o(1 + \varepsilon_A) - P]$ is negative, depending on the stoichiometry and feed composition (i.e. on the value of ε_A). In such cases, Eq. (2.59) cannot be used because it would involve the logarithm of a negative number. By following a similar procedure, described by Eqs. (2.46)–(2.50), the following equation can be derived for these cases:

$$\ln\left(-\frac{dP}{dt}\right) = \ln k^* + n\ln[P - P_o(1 + \varepsilon_A)] \quad \text{For } \varepsilon_A < 0 \tag{2.60}$$

Example 2.5   Solve Example 2.4 using the differential method with data of total pressure.

Solution

From Example 2.4, the value of ε_A is –0.5; thus, Eq. (2.60) must be used. To evaluate the derivative (dP/dt), the method of finite differences will be used, since in this case Δt is the same for all the data, $\Delta t = 100$ sec.

$$\left(\frac{dP}{dt}\right)_0 = \left(\frac{1}{2\Delta t}\right)(-3P_o + 4P_1 - P_2) = -0.1385$$

$$\left(\frac{dP}{dt}\right)_1 = \left(\frac{1}{2\Delta t}\right)(P_2 - P_o) = -0.0935$$

$$\left(\frac{dP}{dt}\right)_2 = \left(\frac{1}{2\Delta t}\right)(P_3 - P_1) = -0.0535$$

$$\left(\frac{dP}{dt}\right)_3 = \left(\frac{1}{2\Delta t}\right)(P_4 - P_2) = -0.0365$$

$$\left(\frac{dP}{dt}\right)_4 = \left(\frac{1}{2\Delta t}\right)(P_5 - P_3) = -0.028$$

$$\left(\frac{dP}{dt}\right)_5 = \left(\frac{1}{2\Delta t}\right)(P_3 - 4P_4 + 3P_5) = -0.010$$

To apply Eq. (2.60), $x = ln[P - P_o(1 + \varepsilon_A)]$ versus $y = ln(-dP/dt)$ are plotted, which gives the following results (Figure 2.17):

Slope	$= n$	$= 2.1253$
Intercept	$= ln\,k$	$= -9.6851$
Correlation coefficient	$= r$	$= 0.9566$

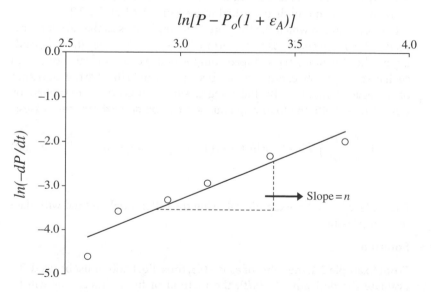

Figure 2.17 Application of the differential method with data of total pressure for Example 2.5.

Therefore, the calculated reaction order is 2.1253. The real reaction order is approximated to an entire value, which is $n = 2$, and with it the value of the reaction rate coefficient is calculated as done in Example 2.4.

2.4   Method of the Half-Life Time

The half-life time ($t_{1/2}$) is the time necessary for the initial concentration to reduce to half of its value. To apply this method, a series of data of $t_{1/2}$ versus C_{Ao} are required (Figure 2.18).

The procedure to determine the reaction order and the reaction rate coefficient is similar to that of the integral method. The only difference is that, when integrating the reaction rate equation, the limits of the integral are from $t = 0$ and C_{Ao}, to $t = t_{1/2}$ and $C_A = C_{Ao}/2$.

Thus, starting with the reaction rate equation for irreversible reactions of one component at constant density, Eq. (2.3):

$$-\frac{dC_A}{dt} = kC_A^n$$

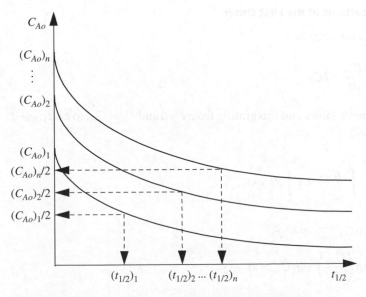

Figure 2.18 Data required to use the method of half-life time.

2.4.1 Reactions of Zero Order

For $n = 0$, Eq. (2.3) is:

$$-\frac{dC_A}{dt} = k$$

Separating variables and integrating from $t = 0$ and $C_A = C_{Ao}$ to $t = t_{1/2}$ and $C_A = C_{Ao}/2$:

$$-\int_{C_{Ao}}^{C_{Ao}/2} dC_A = k \int_{0}^{t_{1/2}} dt = kt_{1/2}$$

$$-C_A\Big|_{C_{Ao}}^{C_{Ao}/2} = kt_{1/2}$$

$$-\frac{C_{Ao}}{2} + C_{Ao} = kt_{1/2}$$

$$\frac{C_{Ao}}{2} = kt_{1/2} \tag{2.61}$$

And the value of k is evaluated with:

$$k = \frac{C_{Ao}}{2t_{1/2}} \tag{2.62}$$

2.4.2 Reactions of the First Order

For $n = 1$, Eq. (2.3) is:

$$-\frac{dC_A}{dt} = kC_A$$

Separating variables and integrating from $t = 0$ and $C_A = C_{Ao}$ to $t = t_{1/2}$ and $C_A = C_{Ao}/2$:

$$-\int_{C_{Ao}}^{C_{Ao}/2} \frac{dC_A}{C_A} = k \int_{0}^{t_{1/2}} dt = kt_{1/2}$$

$$-\ln C_A\Big|_{C_{Ao}}^{C_{Ao}/2} = kt_{1/2}$$

$$-\ln\left(\frac{C_{Ao}}{2}\right) + \ln C_{Ao} = kt_{1/2}$$

$$\ln(2) = kt_{1/2} \tag{2.63}$$

And the value of k is evaluated with:

$$k = \frac{\ln(2)}{t_{1/2}} \qquad (2.64)$$

2.4.3 Reaction of the Second Order

For $n = 2$, Eq. (2.3) is:

$$-\frac{dC_A}{dt} = kC_A^2$$

Separating variables and integrating from $t = 0$ and $C_A = C_{Ao}$ to $t = t_{1/2}$ and $C_A = C_{Ao}/2$:

$$-\int_{C_{Ao}}^{C_{Ao}/2} \frac{dC_A}{C_A^2} = k\int_0^{t_{1/2}} dt = kt_{1/2}$$

$$\left.\frac{1}{C_A}\right|_{C_{Ao}}^{C_{Ao}/2} = kt_{1/2}$$

$$\frac{2}{C_{Ao}} - \frac{1}{C_{Ao}} = kt_{1/2}$$

$$\frac{1}{C_{Ao}} = kt_{1/2} \qquad (2.65)$$

And the value of k is evaluated with:

$$k = \frac{1}{C_{Ao}t_{1/2}} \qquad (2.66)$$

2.4.4 Reaction of the nth Order

Separating variables in Eq. (2.3) and integrating from $t = 0$ and $C_A = C_{Ao}$ to $t = t_{1/2}$ and $C_A = C_{Ao}/2$:

$$-\int_{C_{Ao}}^{C_{Ao}/2} \frac{dC_A}{C_A^n} = k\int_0^{t_{1/2}} dt = kt_{1/2}$$

$$\left.\frac{C_A^{1-n}}{n-1}\right|_{C_{Ao}}^{C_{Ao}/2} = kt_{1/2}$$

$$\frac{1}{n-1}\left[\left(\frac{C_{Ao}}{2}\right)^{1-n} - C_{Ao}^{1-n}\right] = kt_{1/2}$$

$$C_{Ao}^{1-n}(2^{n-1}-1) = k(n-1)t_{1/2} \qquad (2.67)$$

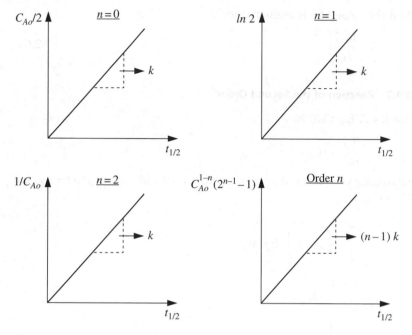

Figure 2.19 Method of half-life time for irreversible reactions of one component.

And the value of k is evaluated with:

$$k = \frac{(2^{n-1}-1)\, C_{Ao}^{1-n}}{(n-1)t_{1/2}} \quad \text{For } n \neq 1 \tag{2.68}$$

Figures 2.19 shows Eqs. (2.61), (2.63), (2.65) and (2.67) in graphical form.

Example 2.6   The thermal decomposition of nitrous oxide (N_2O) was studied at 1030 K, and the data reported in Table 2.12 of $t_{1/2}$ were obtained at different initial partial pressures of N_2O. Find the kinetic model.

Solution

The initial concentration of N_2O for $t_{1/2} = 860\,\text{sec}$ and $(p_{N2O})_o = 82.5\,\text{mmHg}$ is evaluated with:

$$C_{Ao} = \frac{p_{Ao}}{RT} = \frac{(82.5\,mmHg)}{\left(62.36\dfrac{mmHg\, lt}{gmolK}\right)(1030)K} = 1.284 \times 10^{-3}\frac{gmol}{lt}$$

Table 2.12 Data and results of Example 2.6.

$(p_{N2O})_o$ (mmHg)	$t_{1/2}$ (sec)	$C_{Ao} \times 10^3$ (gmol/lt)	$k \times 10^3$ ($n = 1$)	k ($n = 2$)
82.5	860	1.284	0.806	0.9056
139	470	2.164	1.475	0.9832
296	255	4.608	2.718	0.8510
360	212	5.605	3.270	0.8416
Average				0.8954

The integrated equations for the first and second orders of reaction are:
Eq. (2.64) for $n = 1$ is:

$$k = \frac{\ln(2)}{t_{1/2}} = \frac{\ln(2)}{860} = 0.806 \times 10^{-3} \, \text{sec}^{-1}$$

Eq. (2.64) for $n = 2$ is:

$$k = \frac{1}{C_{Ao}t_{1/2}} = \frac{1}{(1.284 \times 10^{-3})(860)} = 0.9056 \frac{lt}{gmol \; sec}$$

The complete results are summarized in Table 2.12. It is observed that for the first order of reaction, the value of k has an increasing tendency, while the second order of reaction accurately fits the experimental data, with an average value of $k = 0.8954$ lt/gmol sec.

It is important to mention that for reactions following the first order, the half-life time is the same to calculate k; in other words, the initial concentration does not have an effect on $n = 1$.

Finally, the kinetic model is:

$$(-r_A) = 0.8954 (lt/gmol \; sec) C_A^2$$

2.4.5 Direct Method to Calculate k and n with Data of $t_{1/2}$

A direct approach to calculate n and k from data of half-life time is by linearization of Eq. (2.68) with logarithms:

$$t_{1/2} = \frac{(2^{n-1} - 1) \, C_{Ao}^{1-n}}{(n-1)k} \qquad (2.69)$$

$$\ln t_{1/2} = \ln \left[\frac{(2^{n-1} - 1)}{(n-1)k} \right] + (1-n) \ln C_{Ao} \qquad (2.70)$$

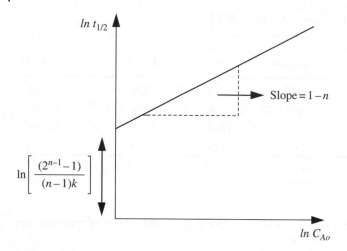

Figure 2.20 Direct method to calculate n and k with data of half-life time.

When plotting Eq. (2.70), a straight line is obtained ($x = lnC_{Ao}$ vs. $y = lnt_{1/2}$). With the slope $(1 - n)$ the reaction order is calculated, and with the intercept the value of k, as illustrated in Figure 2.20.

Example 2.7 Solve Example 2.6 by using the direct method to calculate k and n with data of $t_{1/2}$.

Solution

The data of Example 2.6 are used with $x = ln(C_{Ao})$ versus $y = ln(t_{1/2})$. Using Eq. (2.70), the following results are obtained (Figure 2.21):

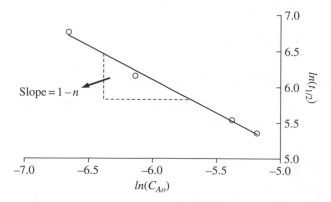

Figure 2.21 Application of the direct method with data of half-life time for Example 2.7.

Slope $\qquad = 1 - n \qquad = -0.9278$

Intercept $\qquad = 0.5342$

Correlation coefficient $\quad = r \qquad = 0.9967$

Therefore, $n = 1.9278$. The entire value of the reaction order is 2, and the reaction rate coefficient is evaluated with the integrated equations used in Example 2.6.

2.4.6 Extension of the Method of Half-Life Time ($t_{1/2}$) to Any Fractional Life Time ($t_{1/m}$)

The method of half-life time can be extended to any fractional life time of the limiting reactant. For instance, $t_{1/3}$ is the time necessary for the initial concentration to reduce to one-third of its value. In general, $t_{1/m}$ is the time necessary for the initial concentration to reduce to $1/m$ of its value.

The integrated equations for any fractional life time are similar to those for half-life time:

For order n ($n \neq 1$):

$$k = \frac{(m^{n-1} - 1)\, C_{Ao}^{1-n}}{(n-1) t_{1/m}} \tag{2.71}$$

For $n = 1$:

$$k = \frac{\ln(m)}{t_{1/m}} \tag{2.72}$$

2.4.7 Calculation of Activation Energy with Data of Half-Life Time

With the data of $t_{1/2}$ or any other fractional life time ($t_{1/m}$), it is possible to determine the activation energy of a reaction if the data of time are reported at different temperatures. The procedure is the following:

Substituting the Arrhenius equation (Eq. 1.91) into Eq. (2.68):

$$t_{1/2} = \frac{(2^{n-1} - 1)\, C_{Ao}^{1-n}}{(n-1) A e^{-E_A/{}_R T}}$$

which can also be written as:

$$t_{1/2} = \frac{(2^{n-1} - 1)\, C_{Ao}^{1-n}}{(n-1) A} \left(e^{E_A/{}_R T} \right)$$

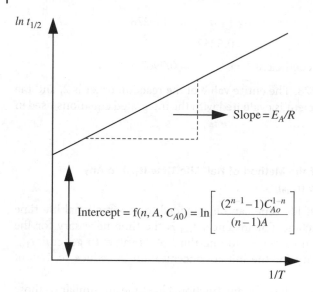

Figure 2.22 Calculation of E_A with data of $t_{1/2}$.

linearization with logarithms gives:

$$\ln t_{1/2} = \ln\left[\frac{(2^{n-1}-1)\,C_{Ao}^{1-n}}{(n-1)A}\right] + \left(\frac{E_A}{R}\right)\frac{1}{T} \tag{2.73}$$

With the slope of this linear equation, the activation energy can be calculated. The intercept is useless because it is a function of the reaction order n and frequency factor A, as illustrated in Figure 2.22. The value of A can only be calculated if the reaction order has been determined with another method.

Example 2.8 The decomposition of N_2O_5 was carried out at different temperatures in an isothermal reactor at constant density. The initial concentration of N_2O_5 was the same for all the experiments. The experimental results are reported in Table 2.13. Calculate the activation energy.

Solution

Using Eq. (2.73) with $x = 1/T$ (T is temperature in K) versus $y = \ln(t_{1/2})$, Figure 2.23 can be prepared. It is seen that the experimental data fit a straight line, from which the following values are obtained:

Slope $= E_A/R$ $= 12458.86$

Intercept $= -31.88$

Correlation coefficient $= r$ $= 0.9999$

Table 2.13 Data and results of Example 2.8.

Temperature (°C)	$t_{1/2}$ (sec)	Temperature (K)	$x = 1/T \times 10^3$	$y = ln(t_{1/2})$
50	780	323.15	3.095	6.659
100	4.6	373.15	2.680	1.526
150	8.8×10^{-2}	423.15	2.363	−2.430
200	3.9×10^{-3}	473.15	2.113	−5.547
300	3.9×10^{-5}	573.15	1.745	−10.152

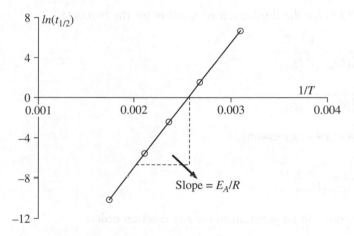

Figure 2.23 Calculation of E_A for Example 2.8.

Therefore, the value of the activation energy is:

$$E_A = 12548.86(R) = 12548.86(1.987) = 24755.8 \, Cal/gmol$$

2.4.8 Some Observations of the Method of Half-Life Time

2.4.8.1 Calculation of *n* with Two Data of $t_{1/2}$ Measured with Different C_{Ao}

If at the same temperature, the half-life time is measured with two experiments using two different initial concentrations of the limiting reactant, $(C_{Ao})_1$ and $(C_{Ao})_2$, it is possible the calculate the reaction order.

For instance, using Eq. (2.65) for the second order of reaction for the two data:

$$k = \frac{1}{(C_{Ao})_1 (t_{1/2})_1}$$

$$k = \frac{1}{(C_{Ao})_2 (t_{1/2})_2}$$

Equaling these two expressions:

$$\frac{(t_{1/2})_1}{(t_{1/2})_2} = \frac{(C_{Ao})_2}{(C_{Ao})_1}$$

Using Eq. (2.66) for the third order of reaction for the two data:

$$k = \frac{1}{[(C_{Ao})_1]^2 (t_{1/2})_1}$$

$$k = \frac{1}{[(C_{Ao})_2]^2 (t_{1/2})_2}$$

Equaling these two expressions:

$$\frac{(t_{1/2})_1}{(t_{1/2})_2} = \left[\frac{(C_{Ao})_2}{(C_{Ao})_1}\right]^2$$

These equations can be generalized for any reaction order:

$$\frac{(t_{1/2})_1}{(t_{1/2})_2} = \left[\frac{(C_{Ao})_2}{(C_{Ao})_1}\right]^{n-1}$$

The linearization by logarithms gives:

$$\ln\left[\frac{(t_{1/2})_1}{(t_{1/2})_2}\right] = (n-1)\ln\left[\frac{(C_{Ao})_2}{(C_{Ao})_1}\right] \tag{2.74}$$

And, finally, the reaction order is calculated with:

$$n = \frac{\ln\left[\dfrac{(t_{1/2})_1}{(t_{1/2})_2}\right]}{\ln\left[\dfrac{(C_{Ao})_2}{(C_{Ao})_1}\right]} + 1 \tag{2.75}$$

2.4.8.2   Generalization of the Method of Half-Life Time for Any Reaction Order

In general, the half-life time of a reaction of order n is related to the initial concentration of the limiting reactant by means of the following relationship:

$$t_{1/2} \propto \frac{1}{C_{Ao}^{n-1}}$$

$$t_{1/2} = \frac{c_{1/2}}{C_{Ao}^{n-1}} = c_{1/2} C_{Ao}^{1-n} \qquad (2.76)$$

where $c_{1/2}$ is a proportionality constant.

For instance, for $n = 0$, Eq. (2.62) is:

$$t_{1/2} = \frac{C_{Ao}}{2k} \quad \text{and} \quad c_{1/2} = 1/2$$

For $n = 1$, Eq. (2.64) is:

$$t_{1/2} = \frac{\ln(2)}{k} \quad \text{and} \quad c_{1/2} = ln(2)$$

For $n = 2$, Eq. (2.65) is:

$$t_{1/2} = \frac{1}{kC_{Ao}} \quad \text{and} \quad c_{1/2} = 1$$

Applying logarithms to Eq. (2.76):

$$\ln t_{1/2} = (1-n)\ln C_{Ao} + \ln(c_{1/2}) \qquad (2.77)$$

Eq. (2.77) is the general form of the method of half-life time. The reaction order is evaluated with the slope $(1 - n)$, and, depending on this reaction order and the intercept ($ln\ c_{1/2}$), the reaction rate coefficient is also calculated.

3

Irreversible Reactions with Two or Three Components

3.1 Irreversible Reactions with Two Components

Irreversible reactions with two components have the following general form:

$$aA + bB \rightarrow Products \tag{3.1}$$

with the following kinetic equation:

$$(-r_A) = k\, C_A{}^{\alpha} C_B{}^{\beta} \tag{3.2}$$

If $\alpha \neq a$ or $\beta \neq b$, the reaction is non-elemental and the kinetic model is represented by Eq. (3.2). If $\alpha = a$ and $\beta = b$, the reaction is elemental and Eq. (3.2) can be written as:

$$(-r_A) = k\, C_A{}^{a} C_B{}^{b} \tag{3.3}$$

Table 3.1 presents some typical reactions between two components.

The methods for the treatment of kinetic data for these reactions are: the integral method, differential method and initial reaction rates method. The method of half-life can also be used, but it only applies to certain reactions depending on the stoichiometry and feed composition.

3.1.1 Integral Method

The integral method is applied following the general procedure described in Section 2.1. That is, from an integrated kinetic equation for an assumed reaction order, the reaction rate coefficient is calculated by analytic or graphic approaches. If the calculated values of k exhibit good agreement with experimental values, the assumed order is correct; if not, it is necessary to assume another reaction order.

In the case of irreversible reactions between two components, there are some particular cases when using the integral method depending on the

Chemical Reaction Kinetics: Concepts, Methods and Case Studies, First Edition.
Jorge Ancheyta.
© 2017 John Wiley & Sons Ltd. Published 2017 by John Wiley & Sons Ltd.

Table 3.1 Examples of irreversible reactions between two components.

Reaction	Kinetic model	Temperature
$N(CH_3)_3 +$ $CH_3CH_2CH_2Br \rightarrow$ $(CH_3)_3(CH_2CH_2CH_3)N^+ +$ Br^-	$(-r_A) = 1.67 \times 10^{-3} \left(lt\, gmol^{-1}\ sec^{-1}\right) C_A C_B$	139.4 °C
$C_2H_4Br_2 + 3KI \rightarrow C_2H_4 +$ $2KBr + KI_3$	$(-r_A) = 8.47 \times 10^{-2} (m^3 kmol^{-1} K sec^{-1}) C_A C_B$	59.7 °C
$C_6H_5CH_3 + H_2 \rightarrow C_6H_6 +$ CH_4	$(-r_A) = 0.316 (ft^{1.5} lbmol^{0.5} sec^{-1}) C_A C_B^{0.5}$	682 °C
$2NO + O_2 \rightarrow 2NO_2$	$(-r_A) = 8.0 \times 10^{-3} (m^6 kmol^{-2} K sec^{-1}) C_A^2 C_B$	30 °C
$CH_4 + 2S_2 \rightarrow CS_2 + 2H_2S$	$(-r_A) = 1.08 \times 10^{-3} (gmol\ cm^{-3} atm^{-2} hr^{-1}) C_A C_B$	600 °C

feed composition, if the reaction is elemental or non-elemental, and if the reacting system is at constant or variable density.

The typical feed compositions used to determine the reaction kinetics are: stoichiometric, non-stoichiometric or with a reactant in excess.

3.1.1.1 Method of Stoichiometric Feed Composition

In Eq. (3.1), for the feed composition to be stoichiometric, the following condition must be fulfilled: $M_{BA} = b/a$, where M_{BA} is the feed molar ratio of B with respect to A ($M_{BA} = N_{Bo}/N_{Ao} = C_{Bo}/C_{Ao}$).

3.1.1.1.1 Elemental Reactions

If Eq. (3.1) is elemental, the kinetic model is given by Eq. (3.3), and to integrate it C_B needs to be put as a function of C_A with Eqs. (1.43)–(1.44):

$$C_A = C_{Ao}(1 - x_A)$$

$$C_B = C_{Ao}\left(M_{BA} - \frac{b}{a}x_A\right)$$

Since $M_{BA} = b/a$:

$$C_B = \frac{b}{a}C_{Ao}(1 - x_A) = \frac{b}{a}C_A$$

Hence, the kinetic model is:

$$(-r_A) = -\frac{dC_A}{dt} = kC_A^a C_B^b = kC_A^a \left(\frac{b}{a}C_A\right)^b = \left(\frac{b}{a}\right)^b kC_A^{a+b}$$

$$(-r_A) = -\frac{dC_A}{dt} = \left(\frac{b}{a}\right)^b kC_A^n \tag{3.4}$$

$$(-r_A) = -\frac{dC_A}{dt} = k'C_A^n \tag{3.5}$$

where n is the global reaction order of the elemental reaction ($n = a + b$).

Similar to the case of variable density in Section 2.1, the kinetic equation is:

$$\frac{C_{Ao}}{1+\varepsilon_A x_A}\frac{dx_A}{dt} = \frac{\left(b/a\right)^b k\, C_{Ao}^n (1-x_A)^n}{(1+\varepsilon_A x_A)^n}$$

$$\frac{dx_A}{dt} = \frac{k'\, C_{Ao}^{n-1}(1-x_A)^n}{(1+\varepsilon_A x_A)^{n-1}} \tag{3.6}$$

where:

$$k' = \left(\frac{b}{a}\right)^b k \tag{3.7}$$

Eqs. (3.5) and (3.6) are similar to those of the irreversible reactions of one component developed in Section 2.1 (Eqs. 2.2 and 2.5), so that the integrated expressions are also similar, changing only k' by k (Eqs. 2.8, 2.14, 2.22 and 2.29 at constant density, and Eqs. 2.11, 2.18, 2.26 and 2.30 at variable density).

The graphical representation of Eqs. (3.5) and (3.7) is similar to that in Figures (2.1)–(2.4), with the difference that the slope of the straight lines changes by k' according to Eq. (3.7).

By using Eqs. (3.5)–(3.7), the values of n and k can be evaluated. Moreover, if the reaction is elemental, the individual reaction orders are also known, $\alpha = a$ and $\beta = b$.

Particular Cases

Case 1: For the reaction $A + B \rightarrow Products$ ($a = b = 1$), the term $(b/a)^b = 1$, and according to Eq. (3.7), $k' = k$, so that the integrated equations for this reaction are identical to those of Section 2.1.

Case 2: For other reaction stoichiometry different from Case 1, or in general when $b \neq a$, the value of k depends on the value of k' [Eq. (3.7)].

3.1.1.1.2 Non-elemental Reaction

If Eq. (3.1) is non-elemental, the kinetic model is given by Eq. (3.2). And, due to the feed composition being stoichiometric, $C_B = (b/a)C_A$; therefore:

$$(-r_A) = -\frac{dC_A}{dt} = kC_A^\alpha C_B^\beta = kC_A^\alpha \left(\frac{b}{a}C_A\right)^\beta = \left(\frac{b}{a}\right)^\beta kC_A^{\alpha+\beta}$$

$$(-r_A) = -\frac{dC_A}{dt} = \left(\frac{b}{a}\right)^\beta kC_A^n \tag{3.8}$$

where n is the global reaction order of the non-elemental reaction $(n = \alpha + \beta)$.

Eq. (3.8) is similar to Eq. (3.4) with the difference that the term (b/a) is raised to the power β instead of to the power b. Eq. (3.8) can be written as:

$$(-r_A) = -\frac{dC_A}{dt} = k' C_A^n \tag{3.9}$$

And, at variable density:

$$\frac{C_{Ao}}{1 + \varepsilon_A x_A} \frac{dx_A}{dt} = \frac{\left(b/a\right)^\beta k \, C_{Ao}^n (1 - x_A)^n}{(1 + \varepsilon_A x_A)^n}$$

$$\frac{dx_A}{dt} = \frac{k' \, C_{Ao}^{n-1} (1 - x_A)^n}{(1 + \varepsilon_A x_A)^{n-1}} \tag{3.10}$$

where:

$$k' = \left(\frac{b}{a}\right)^\beta k \tag{3.11}$$

Again, the equations described in Section 2.1 for irreversible reactions of one component can be applied by changing k' (Eq. 3.11) to k. The integration of Eqs. (3.9) and (3.10) directly gives the value of n.

Particular Cases

Case 3: For reactions where $a = b$, for instance $A + B \rightarrow Products$ $(a = b = 1)$, Eq. (3.11) is:

$$k' = \left(\frac{b}{a}\right)^\beta k = (1)^\beta k = k$$

And the equations reported in Section 2.1 can be directly applied to calculate n and k. The individual reaction orders α and β cannot be determined.

Case 4: For reactions where $a \neq b$, for instance $A + 2B \rightarrow Products$ $(a = 1, b = 2)$ or $2A + B \rightarrow Products$ $(a = 2, b = 1)$, k cannot be determined with Eq. (3.11), because the value of β is required. The value of n is calculated with the equations of Section 2.1. The values of α and β cannot be determined. One alternative to evaluating k is to assume a value of β and calculate α by difference $(\alpha = n - \beta)$.

Table 3.2 summarizes the equations used with the integral method for stoichiometric feed composition, and elemental and non-elemental reactions.

Table 3.2 Integral method for irreversible reactions with two components with stoichiometric feed composition.

Stoichiometric feed composition: $M_{BA} = b/a$

Elemental reaction: $\alpha = a, \beta = b, n = a + b$
$$k' = \left(\frac{b}{a}\right)^b k$$

Non-elemental reaction: $\alpha \neq a, \beta \neq b, n = \alpha + \beta$
$$k' = \left(\frac{b}{a}\right)^\beta k$$

Stoichiometry	Reacting system	Kinetic model	Integrated equation
$aA + bB \rightarrow Products$	Constant density	$(-r_A) = -\dfrac{dC_A}{dt} = k'C_A^n$ $(-r_A) = C_{Ao}\dfrac{dx_A}{dt} = k'C_{Ao}^n(1-x_A)^n$	$\underline{n = 1:}$ $k' = \dfrac{1}{t}\ln\dfrac{C_{Ao}}{C_A} = \dfrac{1}{t}\ln\dfrac{1}{1-x_A}$ $\underline{n \neq 1:}$ $k' = \dfrac{1}{(n-1)t}\left[C_A^{1-n} - C_{Ao}^{1-n}\right]$ $k' = \dfrac{1}{(n-1)C_{Ao}^{n-1}t}\left[(1-x_A)^{1-n} - 1\right]$
$aA + bB \rightarrow Products$	Variable density	$(-r_A) = -\dfrac{dC_A}{dt} = k'C_A^n$ $(-r_A) = \dfrac{C_{Ao}}{1+\varepsilon_A x_A}\dfrac{dx_A}{dt} = \dfrac{k'C_A^n(1-x_A)^n}{(1+\varepsilon_A x_A)^n}$	$\underline{n = 1:}$ $k' = \dfrac{1}{t}\ln\dfrac{1}{1-x_A} = \dfrac{1}{t}\ln\left(\dfrac{C_{Ao}+\varepsilon_A C_A}{C_A + \varepsilon_A C_A}\right)$ $\underline{n \neq 1:}$ $k' = \dfrac{1}{C_{Ao}^{n-1}t}\displaystyle\int_0^{x_A}\dfrac{(1+\varepsilon_A x_A)^{n-1}}{(1-x_A)^n}$

Example 3.1   The following etherification reaction between cyclohexanol and acetic acid was studied in a batch reaction at constant density and 40 °C:

> *Acetic acid + Cyclohexanol*
>
> $\rightarrow$ *Cyclohexyl acetate + Water* $(A + B \rightarrow R + S)$

Table 3.3 presents the experimental results obtained for equal initial concentrations of acetic acid and cyclohexanol ($C_{Ao} = C_{Bo} = 2.5$ kmol/m³). Find the rate equation for this reaction.

Solution

The ratio of stoichiometric coefficient b/a is equal to one, and the ratio of initial concentrations C_{Bo}/C_{Ao} is also equal to one, so that the feed composition is stoichiometric ($M_{BA} = b/a = 1$).

If the elemental reaction order is assumed, $n = 2$, and using the corresponding integrated equation (Table 3.2) for the first data:

$$k' = \frac{1}{(n-1)t}\left[C_A^{1-n} - C_{Ao}^{1-n}\right] = \frac{1}{t}\left[C_A^{-1} - C_{Ao}^{-1}\right] = \frac{1}{7.2}\left[(2.07)^{-1} - (2.5)^{-1}\right]$$

$$= 1.154 \times 10^{-2} m^3 Kmol^{-1} K sec^{-1}$$

The results for the other data are summarized in Table 3.3. Since for the second order of reaction there is a decreasing tendency in the value of k, this is not the correct reaction order. The next assumption is $n = 3$:

Table 3.3 Data and results of Example 3.1.

Time (Ksec)	C_A (Kmol/m³)	$k \times 10^2$ ($n = 2$)	$k \times 10^3$ ($n = 3$)
7.2	2.070	1.154	5.096
9.0	1.980	1.167	5.282
10.8	1.915	1.131	5.217
12.6	1.860	1.092	5.121
14.4	1.800	1.080	5.161
16.2	1.736	1.087	5.303
18.0	1.692	1.061	5.258
19.8	1.635	1.069	5.406
21.6	1.593	1.054	5.418
25.2	1.520	1.023	5.413
28.8	1.460	0.989	5.367
Average	–	–	5.277×10^{-3}

$$k' = \frac{1}{(n-1)t}\left[C_A^{1-n} - C_{Ao}^{1-n}\right] = \frac{1}{2t}\left[C_A^{-2} - C_{Ao}^{-2}\right]$$

$$= \frac{1}{(2)(7.2)}\left[(2.07)^{-2} - (2.5)^{-2}\right] = 5.096 \times 10^{-3} m^6 Kmol^{-2} K sec^{-1}$$

A third reaction order gives better fit, so that the kinetic model is:

$$(-r_A) = 5.277 \times 10^{-3}\left(m^6 Kmol^{-2} K sec^{-1}\right) C_A^\alpha C_B^\beta$$

$$= 5.277 \times 10^{-3}\left(m^6 Kmol^{-2} K sec^{-1}\right) C_A^3$$

It is worth mentioning that with this method, only the global reaction order can be determined, and the individual reaction orders remain unknown.

3.1.1.2 Method of Non-stoichiometric Feed Composition

In Eq. (3.1), for the feed composition to be non-stoichiometric, the following conditions must be fulfilled: $M_{BA} \neq b/a$. The integrated equations for the case of constant density are given in the remainder of this subsection.

3.1.1.2.1 *Elemental Reactions*

If Eq. (3.1) is elemental, the kinetic model is:

$$(-r_A) = kC_A{}^a C_B{}^b$$

At constant density:

$$-\frac{dC_A}{dt} = C_{Ao}\frac{dx_A}{dt}$$

$$C_A = C_{Ao}(1 - x_A)$$

$$C_B = C_{Ao}\left(M_{BA} - \frac{b}{a}x_A\right)$$

Therefore:

$$C_{Ao}\frac{dx_A}{dt} = kC_{Ao}^a(1-x_A)^a C_{Ao}^b\left(M_{BA} - \frac{b}{a}x_A\right)^b$$

$$= kC_{Ao}^n(1-x_A)^a\left(M_{BA} - \frac{b}{a}x_A\right)^b$$

$$\frac{dx_A}{dt} = kC_{Ao}^{n-1}(1-x_A)^a\left(M_{BA} - \frac{b}{a}x_A\right)^b \tag{3.12}$$

where n is the global order of reaction ($n = a + b$). Separating variables:

$$\int_0^{x_A} \frac{dx_A}{(1-x_A)^a \left(M_{BA} - \frac{b}{a}x_A\right)^b} = kC_{Ao}^{n-1} \int_0^t dt = kC_{Ao}^{n-1}t \tag{3.13}$$

For a defined stoichiometric reaction, a, b and n are known, and Eq. (3.13) can be integrated analytically.

Particular Cases

Case 5: For the reaction $A + B \rightarrow Products$, $a = b = 1$ and $n = a + b = 1 + 1 = 2$. Eq. (3.12) is:

$$\int_0^{x_A} \frac{dx_A}{(1-x_A)(M_{BA} - x_A)} = kC_{Ao}t$$

which can be integrated by the method of partial fractions:

$$\int_0^{x_A} \frac{1}{(1-x_A)(M_{BA}-x_A)}dx_A = \int_0^{x_A} \frac{A}{1-x_A}dx_A + \int_0^{x_A} \frac{B}{M_{BA}-x_A}dx_A$$

$$= -A\ln(1-x_A)\big|_0^{x_A} - B\ln(M_{BA}-x_A)\big|_0^{x_A}$$

$$\int_0^{x_A} \frac{1}{(1-x_A)(M_{BA}-x_A)}dx_A = -A\ln(1-x_A) - B\ln(M_{BA}-x_A) + B\ln(M_{BA}) \tag{3.14}$$

To evaluate the constants A and B, the following equality is used:

$$\frac{1}{(1-x_A)(M_{BA}-x_A)} = \frac{A}{1-x_A} + \frac{B}{M_{BA}-x_A} = \frac{A(M_{BA}-x_A) + B(1-x_A)}{(1-x_A)(M_{BA}-x_A)}$$

Comparing the terms, it is observed that the denominator is the same on both sides of the equation, that is, $(1-x_A)(M_{BA}-x_A)$, so that the numerator must also be the same to maintain the equality, so that:

$$A(M_{BA}-x_A) + B(1-x_A) = 1$$
$$AM_{BA} - Ax_A + B - Bx_A = x_A(-A-B) + (AM_{BA}+B) = 1$$

This equation can also be written in the following manner:

$$x_A(-A-B) + (AM_{BA}+B) = x_A(0) + 1$$

And, by comparison of terms:

$$-A - B = 0$$
$$AM_{BA} + B = 1$$

Solving this system of two equations with two unknowns, the following values of the constants A and B are obtained:

$$A = \frac{1}{M_{BA} - 1}$$

$$B = -A$$

So that Eq. (3.14) is:

$$\int_0^{x_A} \frac{dx_A}{(1 - x_A)(M_{BA} - x_A)} = -A\ln(1 - x_A) + A\ln(M_{BA} - x_A) - A\ln(M_{BA})$$

$$= A\ln\left[\frac{M_{BA} - x_A}{M_{BA}(1 - x_A)}\right]$$

And the kinetic model is:

$$\ln\left[\frac{M_{BA} - x_A}{M_{BA}(1 - x_A)}\right] = (M_{BA} - 1)kC_{Ao}t \quad \text{for } M_{BA} \neq 1 \tag{3.15}$$

If M_{BA} were equal to one, the feed composition would be stoichiometric and the equations of Section 3.1.1 would be used.

Case 6: For the reaction: $A + 2B \rightarrow Products$, $a = 1$, $b = 2$ and $n = a + b = 1 + 2 = 3$. Eq. (3.12) is:

$$\int_0^{x_A} \frac{dx_A}{(1 - x_A)(M_{BA} - 2x_A)^2} = kC_{Ao}^2 t$$

Integrating by the partial fractions method:

$$\int_0^{x_A} \frac{dx_A}{(1 - x_A)(M_{BA} - 2x_A)^2} = \int_0^{x_A} \frac{A}{1 - x_A}dx_A + \int_0^{x_A} \frac{B}{M_{BA} - 2x_A}dx_A$$

$$+ \int_0^{x_A} \frac{C}{(M_{BA} - 2x_A)^2}dx_A$$

$$\int_0^{x_A} \frac{dx_A}{(1 - x_A)(M_{BA} - 2x_A)^2} = -A\ln(1 - x_A) + \frac{B}{2}\ln\left(\frac{M_{BA}}{M_{BA} - 2x_A}\right)$$

$$+ C\left[\frac{x_A}{M_{BA}(M_{BA} - 2x_A)}\right] \tag{3.16}$$

Evaluating the constants A and B in a similar manner as in the previous case:

$$A = \frac{1}{(M_{BA} - 2)^2}$$

$$B = -2A$$

$$C = -2A(2 - M_{BA})$$

Therefore, Eq. (3.16) is:

$$\int_0^{x_A} \frac{dx_A}{(1-x_A)(M_{BA} - 2x_A)^2} = -A\ln(1-x_A) - A\ln\left(\frac{M_{BA}}{M_{BA} - 2x_A}\right)$$

$$+ \frac{2A(2 - M_{BA})x_A}{M_{BA}(M_{BA} - 2x_A)}$$

And the kinetic model is:

$$\ln\left[\frac{M_{BA} - 2x_A}{M_{BA}(1 - x_A)}\right] + \frac{(4 - 2M_{BA})x_A}{M_{BA}(M_{BA} - 2x_A)} = (M_{BA} - 2)^2 kC_{Ao}^2 t \quad \text{for } M_{BA} \neq 2$$

$$(3.17)$$

Case 7: For the reaction: $2A + B \rightarrow Products$, $a = 2$, $b = 1$ and $n = a + b = 2 + 1 = 3$. Eq. (3.13) is:

$$\int_0^{x_A} \frac{dx_A}{(1-x_A)^2 \left(M_{BA} - \frac{1}{2}x_A\right)} = kC_{Ao}^2 t$$

By means of integration by the method of partial fractions, the following values of the constants are obtained:

$$A = -\frac{1}{(2M_{BA} - 1)^2}$$

$$B = -(2M_{BA} - 1)A$$

$$C = -\frac{A}{2}$$

And the integrated equation is:

$$\ln\left[\frac{M_{BA}(1 - x_A)}{M_{BA} - \frac{1}{2}x_A}\right] + (2M_{BA} - 1)\left(\frac{x_A}{1 - x_A}\right)$$

$$= \frac{(M_{BA} - 2)^2}{2} kC_{Ao}^2 t \quad \text{for } M_{BA} \neq 1/2 \qquad (3.18)$$

3.1.1.2.2 Non-elemental Reactions

If Eq. (3.1) is non-elemental, the kinetic model is:

$$(-r_A) = k\, C_A{}^{\alpha} C_B{}^{\beta}$$

Thus, in Eq. (3.12), a is changed by α, and b is changed by β. Also, the order of the non-elemental reaction is $n = \alpha + \beta$:

$$\frac{dx_A}{dt} = kC_{Ao}^{n-1}(1-x_A)^{\alpha}\left(M_{BA} - \frac{b}{a}x_A\right)^{\beta}$$

Separating variables:

$$\int_0^{x_A} \frac{dx_A}{(1-x_A)^{\alpha}\left(M_{BA} - \frac{b}{a}x_A\right)^{\beta}} = kC_{Ao}^{n-1}\int_0^t dt = kC_{Ao}^{n-1}t \tag{3.19}$$

The integration of this equation requires knowing the stoichiometry of the reaction, so that the ratio b/a is known. Then, the individual orders of the non-elemental reaction need to be assumed (α and β). Therefore, for a given stoichiometry there is more than one integrated equation, depending on the assumed reaction order, so that it is difficult to determine a general kinetic equation.

Particular Cases
Applying Eq. (3.19):

Case 8: For the reaction: $aA + B \rightarrow Products$. If $\alpha = 1$ and $\beta = 2$, the integrated equation is:

$$\ln\left[\frac{M_{BA} - \frac{1}{a}x_A}{M_{BA}(1-x_A)}\right] + \frac{(1-aM_{BA})}{a^2 M_{BA}}\left(\frac{x_A}{M_{BA} - \frac{1}{a}x_A}\right)$$

$$= \left(\frac{aM_{BA}-1}{a}\right)^2 kC_{Ao}^2 t \quad \text{for } M_{BA} \neq 1/a \tag{3.20}$$

Case 9: For the reaction: $A + bB \rightarrow Products$. If $\alpha = 2$ and $\beta = 1$, the integrated equation is:

$$\ln\left[\frac{M_{BA}(1-x_A)}{M_{BA} - bx_A}\right] + \frac{(M_{BA}-b)}{b}\left(\frac{x_A}{1-x_A}\right) = \frac{(M_{BA}-b)^2}{b}kC_{Ao}^2 t$$

$$\text{for } M_{BA} \neq 1/b$$

$$\tag{3.21}$$

Case 10: For the reaction: $aA + bB \rightarrow Products$. If $\alpha = 1$ and $\beta = 1$, the integrated equation is:

$$\ln\left[\frac{M_{BA} - \frac{b}{a}x_A}{M_{BA}(1 - x_A)}\right] = \left(M_{BA} - \frac{b}{a}\right)kC_{Ao}t \quad \text{for } M_{BA} \neq b/a \qquad (3.22)$$

For elemental reactions at variable density, the equation to integrate is:

$$\int_0^{x_A} \frac{(1 + \varepsilon_A x_A)^{n-1} dx_A}{(1 - x_A)^a\left(M_{BA} - \frac{b}{a}x_A\right)^b} dx_A = kC_{Ao}^{n-1}\int_0^t dt = kC_{Ao}^{n-1}t \qquad (3.23)$$

For non-elemental reactions at variable density, the equation to integrate is:

$$\int_0^{x_A} \frac{(1 + \varepsilon_A x_A)^{n-1} dx_A}{(1 - x_A)^a\left(M_{BA} - \frac{b}{a}x_A\right)^\beta} dx_A = kC_{Ao}^{n-1}\int_0^t dt = kC_{Ao}^{n-1}t \qquad (3.24)$$

The analytical integration of Eqs. (3.23) and (3.24) is not easy. For elemental reactions (Eq. 3.23), the integration becomes easier if the stoichiometry of the reaction is known, since the individual reaction orders would be known ($\alpha = a$, $\beta = b$). For instance, for the reaction: $A + B \rightarrow Products$. If $a = 1$, $b = 1$ and $n = a + b = 1 + 1 = 2$, Eq. (3.23) is:

$$\int_0^{x_A} \frac{1 + \varepsilon_A x_A}{(1 - x_A)(M_{BA} - x_A)} dx_A = kC_{Ao}t$$

Its solution by partial fractions is:

$$(1 + \varepsilon_A)\ln(1 - x_A) + \ln\left[\frac{M_{BA}}{M_{BA} - x_A}\right] = (1 - M_{BA})kC_{Ao}t \quad \text{for } M_{BA} \neq 1 \qquad (3.25)$$

The integration for another stoichiometry of reaction becomes more complicated to solve analytically, and numerical integration is necessary. Table 3.4 summarizes the equations used for the integral method for non-stoichiometric feed composition for elemental and non-elemental reactions.

Table 3.4 Integral method for reactions between two components with non-stoichiometry feed composition.

Stoichiometry	Kinetic model	Integrated equation	Conditions
Elemental reactions			
$A+B \rightarrow Products$	$(-r_A) = k\,C_A C_B$	$\ln\left[\dfrac{M_{BA}-x_A}{M_{BA}(1-x_A)}\right] = (M_{BA}-1)kC_{Ao}t$	$\alpha=1,\ \beta=1,\ n=2$ $M_{BA}\neq 1$
$A+2B \rightarrow Products$	$(-r_A) = k\,C_A C_B^2$	$\ln\left[\dfrac{M_{BA}-2x_A}{M_{BA}(1-x_A)}\right] + \dfrac{(4-2M_{BA})x_A}{M_{BA}(M_{BA}-2x_A)} = (M_{BA}-2)^2\kappa C_{Ao}^2 t$	$\alpha=1,\ \beta=2,\ n=3$ $M_{BA}\neq 2$
$2A+B \rightarrow Products$	$(-r_A) = k\,C_A^2 C_B$	$\ln\left[\dfrac{M_{BA}(1-x_A)}{M_{BA}-\frac{1}{2}x_A}\right] + (2M_{BA}-1)\left(\dfrac{x_A}{1-x_A}\right) = \dfrac{(M_{BA}-2)^2}{2}kC_{Ao}^2 t$	$\alpha=2,\ \beta=1,\ n=3$ $M_{BA}\neq 1/2$
Non-elemental reactions			
$aA+B \rightarrow Products$	$(-r_A) = k\,C_A^a C_B$	$\ln\left[\dfrac{M_{BA}-\frac{1}{a}x_A}{M_{BA}(1-x_A)}\right] + \dfrac{(1-aM_{BA})}{a^2 M_{BA}}\left(\dfrac{x_A}{M_{BA}-\frac{1}{a}x_A}\right) = \left(\dfrac{aM_{BA}-1}{a}\right)^2 kC_{Ao}^2 t$	$\alpha=1,\ \beta=2,\ n=3$ $M_{BA}\neq 1/a$
$A+bB \rightarrow Products$	$(-r_A) = k\,C_A^2 C_B$	$\ln\left[\dfrac{M_{BA}(1-x_A)}{M_{BA}-bx_A}\right] + \dfrac{(M_{BA}-b)}{b}\left(\dfrac{x_A}{1-x_A}\right) = \dfrac{(M_{BA}-b)^2}{b}kC_{Ao}^2 t$	$\alpha=2,\ \beta=1,\ n=3$ $M_{BA}\neq b$
$aA+bB \rightarrow Products$	$(-r_A) = k\,C_A C_B$	$\ln\left[\dfrac{M_{BA}-\frac{b}{a}x_A}{M_{BA}(1-x_A)}\right] = \left(M_{BA}-\dfrac{b}{a}\right)kC_{Ao}t$	$\alpha=1,\ \beta=1,\ n=2$ $M_{BA}\neq b/a$

Table 3.5 Data and results of Example 3.2.

Time (Ksec)	C_A (Kmol/m³)	x_A	$k \times 10^3$ ($\alpha = 1, \beta = 2, n = 3$)	$k \times 10^2$ ($\alpha = 2, \beta = 1, n = 3$)
1.8	0.885	0.115	1.076	9.092
2.7	0.847	0.153	0.980	8.448
4.5	0.769	0.231	0.940	8.478
7.2	0.671	0.329	0.906	8.717
9.0	0.625	0.375	0.859	8.567
12.6	0.544	0.456	0.805	8.613
15.3	0.500	0.500	0.761	8.499
18.0	0.463	0.537	0.723	8.411
Average	–	–	–	8.603×10^{-3}

Example 3.2 The etherification reaction between cyclohexanol and acetic acid of Example 3.1 was also conducted at 40 °C with different initial concentrations of the reactants ($C_{Ao} = 1$ kmol/m³ and $C_{Bo} = 8$ kmol/m³) obtaining the experimental data reported in Table 3.5. Find the kinetic model.

Solution

From Example 3.1, the global order of reaction is already known, $n = 3$. Therefore, the reaction is non-elemental, since the elemental order is 2. Moreover, the ratio of stoichiometric coefficients $b/a = 1$ and the ratio of initial concentrations $C_{Bo}/C_{Ao} = 8$, so that the feed composition is non-stoichiometric ($M_{BA} = 8$ and $b/a = 1$, i.e. $M_{BA} \neq b/a$).

From Table 3.4, it can be observed that for non-elemental reactions with a global order of 3, there are two options of kinetic models:

$$(-r_A) = k\, C_A C_B^2 \quad \alpha = 1, \beta = 2$$

$$\ln\left[\frac{M_{BA} - \frac{1}{a}x_A}{M_{BA}(1 - x_A)}\right] + \frac{(1 - aM_{BA})}{a^2 M_{BA}}\left(\frac{x_A}{M_{BA} - \frac{1}{a}x_A}\right) = \left(\frac{aM_{BA} - 1}{a}\right)^2 kC_{Ao}^2 t$$

$$(-r_A) = k\, C_A^2 C_B \quad \alpha = 2, \beta = 1$$

$$\ln\left[\frac{M_{BA}(1 - x_A)}{M_{BA} - bx_A}\right] + \frac{(M_{BA} - b)}{b}\left(\frac{x_A}{1 - x_A}\right) = \frac{(M_{BA} - b)^2}{b} kC_{Ao}^2 t$$

For the reaction of this example: $a = 1$, $b = 1$ and $M_{BA} = 8$, so that k is calculated with the following equations:

$$(-r_A) = k\, C_A C_B^2 \quad \alpha = 1, \beta = 2$$

$$k = \frac{1}{49\, C_{Ao}^2 t} \left\{ \ln\left[\frac{8 - x_A}{8(1 - x_A)}\right] - \frac{7}{8}\left(\frac{x_A}{8 - x_A}\right) \right\}$$

$$(-r_A) = k\, C_A^2 C_B \quad \alpha = 2, \beta = 1$$

$$k = \frac{1}{49\, C_{Ao}^2 t} \left\{ \ln\left[\frac{8(1 - x_A)}{8 - x_A}\right] + 7\left(\frac{x_A}{1 - x_A}\right) \right\}$$

where:

$$x_A = \frac{C_{Ao} - C_A}{C_{Ao}}$$

The results using these equations are summarized in Table 3.5. It is observed that the best fit is achieved for $\alpha = 2$ and $\beta = 1$, so that the kinetic model is:

$$(-r_A) = 8.603 \times 10^{-3}\left(m^6 Kmol^{-2} K\sec^{-1}\right) C_A^2 C_B$$

3.1.1.3 Method of a Reactant in Excess

When the feed molar ratio of a reactant with respect to another is much higher than the ratio of stoichiometric coefficients of the reaction, the feed composition is considered to be with a reactant in excess. For instance, in Eq. (3.1):

$$aA + bB \rightarrow Products$$

If $M_{BA} >> b/a$, the reactant B is in excess.

Being in excess, the concentration of this reactant does not change considerably as reaction proceeds, as illustrated in Figure 3.1 (i.e. $C_{Bo} \approx C_B \approx$ constant).

With this consideration, Eq. (3.2) is:

$$(-r_A) = kC_A^\alpha C_B^\beta = kC_A^\alpha C_{Bo}^\beta = \left(kC_{Bo}^\beta\right) C_A^\alpha$$

$$(-r_A) = k_{ex} C_A^\alpha \tag{3.26}$$

where:

$$k_{ex} = kC_{Bo}^\beta \tag{3.27}$$

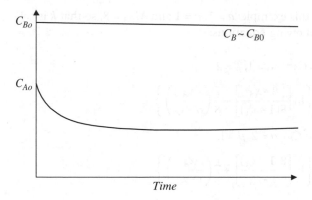

Figure 3.1 Profile of concentrations during a reaction with a reactant in excess.

The solution of this equation is similar to that of the equations of Section 2.1, changing only k_{ex} with k, and n with α:

Constant Density

$$\text{For } \alpha = 1: \quad k_{ex} = \frac{1}{t} \ln \frac{C_{Ao}}{C_A} = \frac{1}{t} \ln \frac{1}{1-x_A} \tag{3.28}$$

$$\text{For } \alpha \neq 1: \quad k_{ex} = \frac{1}{(\alpha-1)t} \left[C_A^{1-\alpha} - C_{Ao}^{1-\alpha} \right]$$

$$= \frac{1}{(\alpha-1) \, C_{Ao}^{\alpha-1} t} \left[\frac{1}{(1-x_A)^{\alpha-1}} - 1 \right] \tag{3.29}$$

Variable Density

$$\text{For } \alpha = 1: \quad k_{ex} = \frac{1}{t} \ln \left(\frac{C_{Ao} + \varepsilon_A C_A}{C_A + \varepsilon_A C_A} \right) = \frac{1}{t} \ln \frac{1}{1-x_A} \tag{3.30}$$

$$\text{For } \alpha^1 1: \quad k_{ex} = \frac{1}{C_{Ao}^{\alpha-1} t} \int_0^{x_A} \frac{(1 + \varepsilon_A x_A)^{\alpha-1}}{(1-x_A)^{\alpha}} dx_A \tag{3.31}$$

With this method, the individual reaction order with respect to the limiting reactant (α) and the reaction rate coefficient (k) is calculated. The reaction order β can be calculated if the global reaction order (n) has been determined by means of another method.

Example 3.3 In Example 3.2 cyclohexanol (A) and acetic acid (B) are fed to the reactor with the following concentrations: $C_{Ao} = 1$ kmol/m^3 and $C_{Bo} = 8$ kmol/m^3. This feed composition, apart from being non-stoichiometric, can be considered as a feed with a reactant in excess (reactant B), because $M_{BA} >> b/a$ ($M_{BA} = 8$ and $b/a = 1$). Hence, the kinetic treatment can also be done by means of the method of a reactant in excess. Table 3.6 reports the experimental data. Find the kinetic model by the method of a reactant in excess.

Solution

From previous examples, the global reaction order is already known, which is $n = 3$. With Eq. (3.29) for $\alpha = 2$, the values of k_{ex} are calculated, which are presented in Table 3.6. For instance, for the first data:

$$k_{ex} = \frac{1}{C_{Ao}^{\alpha-1}t}\left[\frac{1}{(1-x_A)^{\alpha-1}} - 1\right] = \frac{1}{C_{Ao}t}\left[\frac{1}{1-x_A} - 1\right]$$

$$= \frac{1}{(1)(1.8)}\left[\frac{1}{1-0.115} - 1\right] = 7.219 \times 10^{-2}$$

From Eq. (3.27) for $\beta = 1$ and $C_{Bo} = 8$ kmol/m^3:

$$k = \frac{k_{ex}}{C_{Bo}^{\beta}} = \frac{k_{ex}}{C_{Bo}} = \frac{6.712 \times 10^{-2}}{8} = 8.390 \times 10^{-3}$$

Since the values of k are more or less constant, it is assumed that the proposed value of α is correct. Hence, the kinetic model is:

$$(-r_A) = 8.39 \times 10^{-3}\left(m^6 Kmol^{-2} Ksec^{-1}\right)C_A^2 C_B$$

Table 3.6 Data and results of Example 3.3.

Time (Ksec)	C_A (Kmol/m^3)	x_A	$k_{ex} \times 10^2$ ($\alpha = 2$, $\beta = 1$, n = 3)
1.8	0.885	0.115	7.219
2.7	0.847	0.153	6.690
4.5	0.769	0.231	6.675
7.2	0.671	0.329	6.810
9.0	0.625	0.375	6.667
12.6	0.544	0.456	6.653
15.3	0.500	0.500	6.536
18.0	0.463	0.537	6.443
Average	–	–	6.712×10^{-2}

3.1.2 Differential Method

For reactions between two components:

$$aA + bB \longrightarrow Products$$

The differential method considers the linearization of the following equation based on the feed composition:

$$(-r_A) = kC_A^\alpha C_B^\beta \tag{3.32}$$

3.1.2.1 Stoichiometric Feed Composition

If the feed composition is stoichiometric, Eq. (3.32) is:

$$(-r_A) = k'C_A^n$$

where:

For elemental reaction: $k' = \left(\dfrac{b}{a}\right)^b$

For non-elemental reaction: $k' = \left(\dfrac{b}{a}\right)^\beta$

By means of linear regression, the equations of Section 2.2 can be applied to evaluate n and k':

$$\ln\left(-\frac{dC_A}{dt}\right) = \ln k' + n\ln C_A \tag{3.33}$$

$$\ln\left(\frac{dx_A}{dt}\right) = \ln\left(k'C_{Ao}^{n-1}\right) + n\ln(1 - x_A) \tag{3.34}$$

3.1.2.2 Feed Composition with a Reactant in Excess

If in the feed composition there is a reactant in excess (e.g. B), Eq. (3.32) is:

$$(-r_A) = k_{ex}C_A^\alpha$$

where:

$$k_{ex} = kC_{Bo}^\beta$$

And the linear equations are:

$$\ln\left(-\frac{dC_A}{dt}\right) = \ln k_{ex} + \alpha\ln C_A \tag{3.35}$$

$$\ln\left(\frac{dx_A}{dt}\right) = \ln\left(k_{ex}C_{Ao}^{\alpha-1}\right) + \alpha\ln(1 - x_A) \tag{3.36}$$

3.1.2.3 Non-stoichiometric Feed Compositions

If the feed is non-stoichiometric, Eq. (3.32) is used directly in its linear form:

$$\ln\left(-\frac{dC_A}{dt}\right) = \ln k + \alpha \ln C_A + \beta \ln C_B \tag{3.37}$$

Transforming this equation into a linear expression:

$$y = a_0 + a_1 x_1 + a_2 x_2 \tag{3.38}$$

where:

$$a_0 = \ln k$$
$$y = \ln(-dC_A/dt)$$
$$a_1 = \alpha$$
$$x_1 = \ln C_A$$
$$a_2 = \beta$$
$$x_2 = \ln C_B$$

To evaluate the constants a_0, a_1 and a_2, and consequently the kinetic parameters (k, α and β), the following equations can be used:

$$a_0 N + a_1 \sum_{i-1}^{N} x_1 + a_2 \sum_{i=1}^{N} x_2 = \sum_{i=1}^{N} y \tag{3.39}$$

$$a_0 \sum_{i=1}^{N} x_1 + a_1 \sum_{i=1}^{N} x_1^2 + a_2 \sum_{i=1}^{N} x_1 x_2 = \sum_{i=1}^{N} x_1 y \tag{3.40}$$

$$a_0 \sum_{i=1}^{N} x_2 + a_1 \sum_{i=1}^{N} x_1 x_2 + a_2 \sum_{i=1}^{N} x_2^2 = \sum_{i=1}^{N} x_2 y \tag{3.41}$$

where N is the number of experimental data.

The value of $(-dC_A/dt)$ to calculate y in the previous equations is evaluated by means of numerical or graphical differentiation with the methods described in Section 2.2. With this system of three equations with three unknowns, the constants a_0, a_1 and a_2 can be calculated.

Example 3.4 The following liquid phase reaction was studied in a batch reactor:

$$A + B \longrightarrow Products$$

The experimental data reported in Table 3.7 were obtained at 80 °C with the following initial concentrations of reactants and products. Find the rate equation for this reaction.

Table 3.7 Data and results of Example 3.4.

Time (min)	C_A (mol/lt)	x_A	C_B (mol/lt)	$\Delta C_A/\Delta t$	C_{Ap} (mol/lt)	C_{Bp} (mol/lt)
3.10	1.60	0.20	4.60	0.0972	1.46	4.46
5.98	1.32	0.34	4.32	0.0779	1.23	4.23
8.29	1.14	0.43	4.14	0.0466	0.85	3.85
20.74	0.56	0.72	3.56	0.0175	0.35	3.35
27.19	0.40	0.80	3.40	0.0120	0.25	3.25
32.92	0.30	0.85	3.30	0.0048	0.10	1.60
41.25	0.20	0.90	3.20	–	–	–

$$C_{Ao} = 2\,mol/lt$$
$$C_{Bo} = 5\,mol/lt$$
$$C_{Ro} = C_{So} = 0$$

Solution

The feed molar ratio M_{BA} and the ratio of stoichiometric coefficients b/a are:

$$M_{BA} = \frac{C_{Bo}}{C_{Ao}} = \frac{5}{2} = 2.5$$

Since $M_{BA} \neq b/a$, the feed composition is non-stoichiometric. To apply the differential method, the concentrations of B are required, which can be evaluated in the following way. For instance, for the first data:

$$x_A = \frac{C_{Ao} - C_A}{C_{Ao}} = \frac{2.0 - 1.6}{2.0} = 0.20$$
$$C_B = C_{Ao}\left(M_{BA} - \frac{b}{a}x_A\right) = 2.0(2.5 - 0.20) = 4.6\,mol/lt$$

The values of the derivate (dC_A/dt) and average concentrations C_{Ap} and C_{Bp} are also needed. In this case, Δt is not the same for all the data, so the method of finite differences cannot be used; thus, the derivative was estimated as an approximation to $\Delta C_A/\Delta t$. The corresponding values are calculated as:

$$\left(\frac{\Delta C}{\Delta t}\right)_1 = \frac{C_{A1} - C_{A2}}{t_2 - t_1} = \frac{1.60 - 1.32}{5.98 - 3.1} = 0.0972$$
$$(C_{Ap})_1 = \frac{C_{A1} + C_{A2}}{2} = \frac{1.60 + 1.32}{2} = 1.46$$
$$(C_{Bp})_1 = \frac{C_{B1} + C_{B2}}{2} = \frac{4.60 + 4.32}{2} = 4.46$$

Table 3.8 Information required for solving Eqs. (3.39)–(3.41) of Example 3.4.

$\ln(\Delta C_A/\Delta t)$ y	$\ln C_{Ap}$ x_1	$\ln C_{Bp}$ x_2	$x_1\,y$	$x_1\,x_2$	$x_2\,y$	x_1^2	x_2^2
−2.3309	0.3784	1.4951	−0.8820	0.5657	−3.4849	0.1432	2.2353
−2.5523	0.2070	1.4422	−0.5283	0.2985	−3.6809	0.0428	2.0799
−3.0662	−0.1625	1.3481	0.4983	−0.2191	−4.1335	0.0264	1.8174
−4.0456	−1.0498	1.2089	4.2471	−1.2691	−4.8907	1.1021	1.4614
−4.4228	−1.3863	1.1787	6.1313	−1.6340	−5.2132	1.9218	1.3893
−5.3391	−2.3026	0.4700	12.2938	−1.0822	−2.5094	5.3020	0.2209
−21.7569	−4.3158	7.1430	21.7602	−3.3402	−23.9126	8.5383	9.2042

The complete results are summarized in Table 3.7, which are used to calculate the data required to solve Eqs. (3.39)–(3.41) (Table 3.8), which provide the following values:

$$a_0 = -4.2$$
$$a_1 = 1$$
$$a_2 = 1$$

And, finally, the values of kinetic parameters are:

$$\alpha = 1$$
$$\beta = 1$$
$$k = 0.015$$

Thus, the kinetic model is:

$$(-r_A) = 0.015\left(lt\,mol^{-1}\min^{-1}\right)C_A C_B$$

3.1.3 Method of Initial Reaction Rates

If the following kinetic model for irreversible reactions between two components:

$$(-r_A) = kC_A^\alpha C_B^\beta$$

is used at the initial conditions of the reaction, it transforms to:

$$(-r_A)_o = kC_{Ao}^\alpha C_{Bo}^\beta \tag{3.42}$$

This transformation facilitates the evaluation of the kinetic parameters, since experimentally the initial reaction rate can be easily measured for known values of initial concentration of reactants. The method requires two series of experiments.

Experiment 1
The concentration of reactant B is kept constant, and the variation of the concentration of A with respect to time is determined. The initial reaction rate is evaluated with the slope of the curve at zero time, as illustrated in Figure 3.2. Under these conditions. Eq. (3.42) is:

$$C_{Ao} = \text{variable}$$
$$C_{Bo} = \text{constant} \tag{3.43}$$
$$(-r_A)_o = kC_{Ao}^\alpha C_{Bo}^\beta = \left(kC_{Bo}^\beta\right)C_{Ao}^\alpha = k_{v1}C_{Ao}^\alpha$$

Experiment 2
The concentration of reactant A is kept constant, and the variation of the concentration of B with respect to time is determined. The initial rate is evaluated with the slope of the curve at zero time, as illustrated in Figure 3.3. Under these conditions. Eq. (3.42) is:

$$C_{Ao} = \text{constant}$$
$$C_{Bo} = \text{variable} \tag{3.44}$$
$$(-r_A)_o = kC_{Ao}^\alpha C_{Bo}^\beta = \left(kC_{Ao}^\alpha\right)C_{Bo}^\beta = k_{v2}C_{Bo}^\beta$$

From Experiment 1, Eq. (3.43) is transformed into a linear equation, whereby the slope of the straight line gives the value of α and with the intercept the value of k_{v1}.

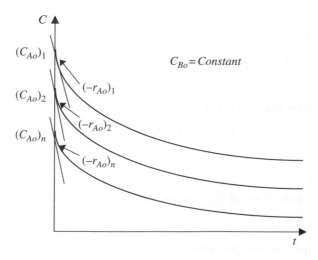

Figure 3.2 Experiment 1 for the method of initial reaction rates.

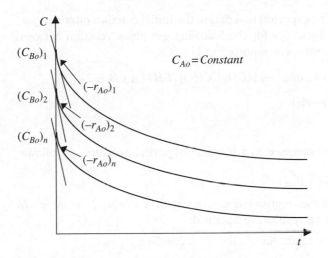

Figure 3.3 Experiment 2 for the method of initial reaction rates.

$$\ln(-r_A)_o = \ln k_{v1} + \alpha \ln C_{Ao} \tag{3.45}$$

With the value of k_{v1}, the reaction rate coefficient for this experiment is evaluated as:

$$k_1 = \frac{k_{v1}}{C_{Bo}^{\beta}} \tag{3.46}$$

From Experiment 2, Eq. (3.44) is transformed into a linear equation, whereby the slope of the straight line gives the value of β and with the intercept the values of k_{v2}.

$$\ln(-r_A)_o = \ln k_{v2} + \beta \ln C_{Bo} \tag{3.47}$$

With the value of k_{v2}, the reaction rate coefficient for this experiment is evaluated as:

$$k_2 = \frac{k_{v2}}{C_{Ao}^{\alpha}} \tag{3.48}$$

Note that to evaluate k_1 the value of β is required, and to evaluate k_2 the value of α is needed, so that the calculation of the kinetic parameters requires the data of the two experiments.

The value of the reaction rate coefficient is determined as the average value of k_1 and k_2:

$$k = \frac{k_1 + k_2}{2} \tag{3.49}$$

Example 3.5   The experimental data of the initial reaction rates and partial pressures of Table 3.9 for the following gas phase reaction between diborane and ketone were reported at 114 °C:

$$Diborane + ketone \rightarrow 2\left((CH_3)_2CHO\right)_2BH \quad (A + B \rightarrow 2R)$$

Find the kinetic model.

Solution

Eq. (3.42) can be expressed as a function of partial pressures as follows:

$$(-r_A)_o = k_p p_{Ao}^{\alpha} p_{Bo}^{\beta}$$

By means of linear regression analysis with $x - ln(p_{Ao})$ versus $y - ln$ $(-r_A)_o$, the following results are obtained:

Slope of the straight line	$= \alpha$	$= 0.9854 \approx 1$
Intercept	$= ln\ k_{v1}$	$= -9.3796 \Rightarrow k_{v1} = 8.4429 \times 10^{-5}$
Correlation coefficient	$= r$	$= 0.9975$

Using the same procedure with $x = ln(p_{Bo})$ versus $y = ln(-r_A)_o$, the following results are obtained:

Slope of the straight line	$= \beta$	$= 0.9957 \approx 1$
Intercept	$= ln\ k_{v2}$	$= -10.2161 \Rightarrow k_{v2} = 3.6577 \times 10^{-5}$
Correlation coefficient	$= r$	$= 0.9959$

Table 3.9 Data and results of Example 3.5.

Run	p_{Ao} (torr)	p_{Bo} (torr)	$(-r_A)_o \times 10^3$ (torr/sec)	$ln(p_{Ao})$	$ln(p_{Bo})$	$ln(-r_A)_o$
1	6	20	0.50	1.7918	–	−7.6009
2	8	20	0.63	2.0794	–	−7.3698
3	10	20	0.83	2.3026	–	−7.0941
4	12	20	1.00	2.4849	–	−6.9078
5	16	20	1.28	2.7726	–	−6.6609
6	10	10	0.33	–	2.3026	−8.0164
7	10	20	0.80	–	2.9957	−7.1309
8	10	40	1.50	–	3.6889	−6.5023
9	10	60	2.21	–	4.0943	−6.1148
10	10	100	3.33	–	4.6052	−5.7048

From Eqs. (3.46) and (3.48) as functions of partial pressures, the values of k_1 and k_2 can be determined, and finally with Eq. (3.49) the reaction rate coefficient k_p is calculated, as follows:

$$k_1 = \frac{k_{v1}}{p_{Bo}^\beta} = \frac{8.4429 \times 10^{-5}}{20} = 4.2215 \times 10^{-6}$$

$$k_2 = \frac{k_{v2}}{p_{Ao}^\alpha} = \frac{3.6577 \times 10^{-5}}{10} = 3.6577 \times 10^{-6}$$

$$k_p = \frac{k_1 + k_2}{2} = \frac{4.2215 \times 10^{-6} + 3.6577 \times 10^{-6}}{2} = 3.9346 \times 10^{-6} torr^{-1}sec^{-1}$$

And the kinetic model is:

$$(-r_A) = 3.9346 \times 10^{-6} (torr^{-1}sec^{-1}) p_A p_B$$

3.2 Irreversible Reactions between Three Components

The irreversible reactions between three components present the following general form:

$$aA + bB + dD \rightarrow Products \tag{3.50}$$

With the following kinetic expression:

$$(-r_A) = kC_A{}^\alpha C_B{}^\beta C_D{}^\gamma \tag{3.51}$$

If $\alpha \neq a$ or $\beta \neq b$ or $\gamma \neq d$, the reaction is non-elemental and the kinetic model is represented by Eq. (3.51). If $\alpha = a$, $\beta = b$ and $\gamma = d$, the reaction is elemental and Eq. (3.51) is transformed to:

$$(-r_A) = kC_A{}^a C_B{}^b C_D{}^d \tag{3.52}$$

Similarly to the case of two components, for irreversible reactions between three components there are different options for the application of integral and differential methods, depending on the feed composition used during the reaction. Some typical cases are described in the remainder of this section.

3.2.1 Case 1: Stoichiometric Feed Composition

If the feed composition is stoichiometric: $M_{BA} = b/a$ and $M_{DA} = d/a$.

Elemental Reactions
For the reaction:

$$aA + bB + dD \rightarrow Products$$

the kinetic model is given by Eq. (3.52):

$$(-r_A) = kC_A{}^a C_B{}^b C_D{}^d$$

which can be put as a function of the concentration of the limiting reactant as follows:

$$(-r_A) = kC_{Ao}^a(1-x_A)^a C_{Ao}^b\left(M_{BA} - \frac{b}{a}x_A\right)^b C_{Ao}^d\left(M_{DA} - \frac{d}{a}x_A\right)^d$$

$$(-r_A) = k\left(\frac{b}{a}\right)^b\left(\frac{d}{a}\right)^d C_{Ao}^n(1-x_A)^n$$

$$(-r_A) = k'C_{Ao}^n(1-x_A)^n = k'C_A^n \qquad (3.53)$$

where n is the global order of the elemental reaction ($n = a+b+d$) and:

$$k' = k\left(\frac{b}{a}\right)^b\left(\frac{d}{a}\right)^d \qquad (3.54)$$

Non-elemental Reaction
For the reaction:

$$aA + bB + dD \rightarrow Products$$

the kinetic model is given by Eq. (3.51):

$$(-r_A) = kC_A^\alpha C_B^\beta C_D^\gamma$$

which can be put as a function of the concentration of the limiting reactant as follows:

$$(-r_A) = kC_{Ao}^\alpha(1-x_A)^\alpha C_{Ao}^\beta\left(M_{BA} - \frac{b}{a}x_A\right)^\beta C_{Ao}^\gamma\left(M_{DA} - \frac{d}{a}x_A\right)^\gamma$$

$$(-r_A) = k\left(\frac{b}{a}\right)^\beta\left(\frac{d}{a}\right)^\gamma C_{Ao}^n(1-x_A)^n$$

$$(-r_A) = k'C_{Ao}^n(1-x_A)^n = k'C_A^n \qquad (3.55)$$

where n is the global order of the non-elemental reaction ($n = \alpha + \beta + \gamma$) and:

$$k' = k\left(\frac{b}{a}\right)^\beta\left(\frac{d}{a}\right)^\gamma \qquad (3.56)$$

With these considerations, when the integral method is used [Eq. (3.53) or Eq. (3.55)], the solution is similar to those of reactions with one component:

For $n = 1$:

$$k' = \frac{1}{t}\ln\frac{C_{Ao}}{C_A} = \frac{1}{t}\ln\frac{1}{1-x_A} \qquad (3.57)$$

For $n \neq 1$:

$$k' = \frac{1}{(n-1)t}\left[C_A^{1-n} - C_{Ao}^{1-n}\right] = \frac{1}{(n-1)\,C_{Ao}^{n-1}t}\left[(1-x_A)^{1-n} - 1\right] \qquad (3.58)$$

The solution by means of the differential method is with the following linear equations:

$$\ln\left(-\frac{dC_A}{dt}\right) = \ln k' + n\ln C_A \qquad (3.59)$$

$$\ln\left(\frac{dx_A}{dt}\right) = \ln\left(k'C_{Ao}^{n-1}\right) + n\ln(1-x_A) \qquad (3.60)$$

The evaluation of the reaction rate coefficient from the value of k' depends on the type of reaction, elemental or non-elemental, according to Eqs. (3.54) and (3.56). In the particular case of $a = b = d$, the reaction rate coefficient k is equal to the constant k'.

3.2.2 Case 2: Non-stoichiometric Feed Composition

If the feed composition is non-stoichiometric: $M_{BA} \neq b/a$ and $M_{DA} \neq d/a$. And the kinetic model is:

$$(-r_A) = kC_{Ao}^{a}(1-x_A)^{a}C_{Ao}^{\beta}\left(M_{BA} - \frac{b}{a}x_A\right)^{\beta} C_{Ao}^{\gamma}\left(M_{DA} - \frac{d}{a}x_A\right)^{\gamma}$$

$$C_{Ao}\frac{dx_A}{dt} = kC_{Ao}^{n}(1-x_A)^{a}\left(M_{BA} - \frac{b}{a}x_A\right)^{\beta}\left(M_{DA} - \frac{d}{a}x_A\right)^{\gamma}$$

$$\int_0^{x_A} \frac{dx_A}{(1-x_A)^{a}\left(M_{BA} - \frac{b}{a}x_A\right)^{\beta}\left(M_{DA} - \frac{d}{a}x_A\right)^{\gamma}} = kC_{Ao}^{n-1}\int_0^{t} dt = kC_{Ao}^{n-1}t$$

$$(3.61)$$

For non-elemental reactions, the integration of Eq. (3.61) requires the supposition of the individual orders of reaction.

For the case of elemental reactions, the integration of Eq. (3.61) depends only on the stoichiometry of the reaction. For instance, for the following reaction:

$$A + B + D \rightarrow Products$$

the kinetic model is:

$$(-r_A) = kC_A C_B C_D$$

The integration of Eq. (3.61) as a function of concentrations results in:

$$k = \frac{1}{t}\left[\frac{1}{(C_{Ao}-C_{Bo})(C_{Ao}-C_{Do})}\ln\frac{C_{Ao}}{C_A} + \frac{1}{(C_{Bo}-C_{Ao})(C_{Bo}-C_{Do})}\right.$$
$$\left.\ln\frac{C_{Bo}}{C_B} + \frac{1}{(C_{Do}-C_{Ao})(C_{Do}-C_{Bo})}\ln\frac{C_{Do}}{C_D}\right]$$

$$(3.62)$$

or:

$$k = \frac{1}{(C_{Ao}-C_{Bo})(C_{Bo}-C_{Do})(C_{Do}-C_{Ao})t}$$
$$\ln\left[\left(\frac{C_A}{C_{Ao}}\right)^{C_{Bo}-C_{Do}}\left(\frac{C_B}{C_{Bo}}\right)^{C_{Do}-C_{Ao}}\left(\frac{C_D}{C_{Do}}\right)^{C_{Ao}-C_{Bo}}\right]$$

$$(3.63)$$

3.2.3 Case 3: Feed Composition with One Reactant in Excess

For the reaction:

$$aA + bB + dD \rightarrow Products$$

With a reactant in excess in the feed, for instance reactant D:

$$M_{DA} >> d/a$$

$$C_{Do} \approx C_D = constant$$

Elemental Reaction
The kinetic model is given by Eq. (3.52):

$$(-r_A) = kC_A{}^a C_B{}^b C_D{}^d$$
$$(-r_A) = kC_A^a C_B^b C_D^d = kC_A^a C_B^b C_{Do}^d = (kC_{Do}^d)C_A^a C_B^b$$
$$(-r_A) = k_{ex}C_A^a C_B^b \qquad (3.64)$$

where:

$$k_{ex} = kC_{Do}^d \qquad (3.65)$$

Non-elemental Reaction
The kinetic model is given by Eq. (3.51):

$$(-r_A) = kC_A{}^\alpha C_B{}^\beta C_D{}^\gamma$$
$$(-r_A) = kC_A^\alpha C_B^\beta C_D^\gamma = kC_A^\alpha C_B^\beta C_{Do}^\gamma = (kC_{Do}^\gamma)C_A^\alpha C_B^\beta$$
$$(-r_A) = k_{ex}C_A^\alpha C_B^\beta \qquad (3.66)$$

where:

$$k_{ex} = kC_{Do}^{\gamma} \tag{3.67}$$

Using the integral method, Eqs. (3.64) and (3.66) are solved in a similar way as those of the case of irreversible reactions with two components with non-stoichiometric feed composition:

$$\int_0^{x_A} \frac{dx_A}{(1-x_A)^{\alpha}\left(M_{BA}-\dfrac{b}{a}x_A\right)^{\beta}} = k_{ex}C_{Ao}^{n-1}\int_0^t dt = k_{ex}C_{Ao}^{n-1}t \tag{3.68}$$

The integration of Eq. (3.68) depends on the type of reaction, elemental or non-elemental, in addition to the stoichiometry of the reaction and the molar feed ratio of B with respect to A (M_{BA}).

For stoichiometric feed composition of B with respect to A (i.e. $M_{BA} = b/a$), the integration of Eq. (3.68) is similar to the equations presented in Table 3.2.

For non-stoichiometric feed composition of B with respect to A (i.e. $M_{BA} \neq b/a$), the equations presented in Table 3.3 are applicable for reactions between three components.

In both cases, the constant k' has to be evaluated; this includes the constant k_{ex}, from which the reaction rate coefficient k can be calculated from Eqs. (3.65) or (3.67).

3.2.4 Case 4: Feed Composition with Two Reactants in Excess

For two reactants in excess:

$$M_{BA} \gg b/a$$
$$C_{Bo} \approx C_B = constant$$
$$M_{DA} \gg d/a$$
$$C_{Do} \approx C_D = constant$$

Elemental Reactions
For the reaction:

$$aA + bB + dD \rightarrow Products$$

the kinetic model is given by Eq. (3.52):

$$(-r_A) = kC_A{}^a C_B{}^b C_D{}^d$$
$$(-r_A) = kC_A^a C_B^b C_D^d = kC_A^a C_{Bo}^b C_{Do}^d = \left(kC_{Bo}^b C_{Do}^d\right)C_A^a$$
$$(-r_A) = k_{ex}C_A^a \tag{3.69}$$

where:

$$k_{ex} = kC_{Bo}^{b} C_{Do}^{d} \qquad (3.70)$$

Non-elemental Reactions
For the reaction:

$$aA + bB + dD \rightarrow Products$$

the kinetic model is given by Eq. (3.51):

$$(-r_A) = kC_A{}^{\alpha} C_B{}^{\beta} C_D{}^{\gamma}$$

$$(-r_A) = kC_A^{\alpha} C_B^{\beta} C_D^{\gamma} = kC_A^{\alpha} C_{Bo}^{\beta} C_{Do}^{\gamma} = \left(kC_{Bo}^{\beta} C_{Do}^{\gamma} \right) C_A^{\alpha}$$

$$(-r_A) = k_{ex} C_A^{\alpha} \qquad (3.71)$$

where:

$$k_{ex} = kC_{Bo}^{\beta} C_{Do}^{\gamma} \qquad (3.72)$$

Using the integral method, Eqs. (3.69) and (3.71) are solved in a similar way as those of the case of irreversible reactions with one component. By this means, the reaction order with respect to A (α) is obtained.

For elemental reactions, the reaction rate coefficient is obtained directly from Eq. (3.70). For non-elemental reactions, the reaction rate coefficient can only be evaluated if the individual orders of reaction with respect to B and D (i.e. β and γ) are known by using Eq. (3.72).

Example 3.6 The following reaction:

$$A + B + D \rightarrow R + S + T$$

was studied at 25 °C by measuring the change of electric resistance with initial concentrations of A, B and D of 0.0045, 0.1 and 0.1 gmol/lt, respectively. The experimental data are reported in Table 3.10. Find the rate equation for this reaction.

Solution

The conversion of A can be evaluated as a function of the electric resistance with Eq. (1.27):

$$x_A = \frac{\Omega - \Omega_o}{\Omega_\infty - \Omega_o}$$

where:

Ω : Electric resistance at time t
Ω_o: Electric resistance at zero time ($t = 0$)
Ω_∞: Electric resistance that does not change with time.

Table 3.10 Data and results of Example 3.6.

Time (min)	Electric resistance (Ω)	x_A	$k_{ex} \times 10^2$ ($\alpha = 1,\ \beta = 1,\ \gamma = 1$)
0	2503	0.0000	–
5	2295	0.2014	4.498
10	2125	0.3659	4.555
15	1980	0.5063	4.706
20	1880	0.6031	4.620
25	1778	0.7018	4.840
30	1719	0.7590	4.743
∞	1470	1.0000	–
Average	–	–	4.615×10^{-2}

From Table 3.10:

Ω_o = 2503
Ω_∞ = 1470

so that the conversion for each time is evaluated with:

$$x_A = \frac{\Omega - 2503}{1470 - 2503}$$

The complete results are summarized in Table 3.10.

The feed molar ratios are:

$$M_{BA} = \frac{C_{Bo}}{C_{Ao}} = \frac{0.1}{0.0045} = 22.22$$

$$M_{DA} = \frac{C_{Do}}{C_{Ao}} = \frac{0.1}{0.0045} = 22.22$$

Since $M_{BA} >> b/a$ and $M_{DA} >> d/a$, the feed composition has two reactants in excess. Assuming that the reaction is elemental ($\alpha = a = 1, \beta = b = 1$, and $\gamma = d = 1$), thus the kinetic model is given by Eq. (3.69):

$$(-r_A) = k_{ex} C_A$$

$$k_{ex} = k C_{Bo} C_{Do}$$

Since $\alpha = 1$, the integration of this equation is:

$$k_{ex} = \frac{1}{t} \ln \frac{1}{1 - x_A}$$

The complete results are presented in Table 3.10.

As can be observed, the values of k_{ex} are similar, with an average value of 4.615×10^{-2}. The value of the reaction rate coefficient is:

$$k = \frac{k_{ex}}{C_{Bo}C_{Do}} = \frac{4.615 \times 10^{-2}}{(0.1)(0.1)} = 4.615$$

And, finally, the kinetic model is:

$$(-r_A) = 4.615 \left(gmol^{-2} lt^2 min^{-1} \right) C_A C_B C_D$$

4

Reversible Reactions

A reversible chemical reaction is that in which the reaction is not completed 100 per cent, that is, there is an equilibrium conversion that corresponds to an infinite reaction time.

By thermodynamics, it is shown that the total Gibbs free energy of a closed system at constant temperature and pressure should diminish during an irreversible reaction, and that the equilibrium condition is achieved when $(dG)^t{}_{T,P} = 0$.

4.1 Reversible Reactions of First Order

This type of reaction has the following stoichiometry and kinetic model:

$$aA \underset{k_2|}{\overset{k_1|}{\rightleftharpoons}} rR \qquad (4.1)$$

$$(-r_A) = k_1 C_A - k_2 C_R \qquad (4.2)$$

where k_1 is the rate coefficient of the direct reaction, and k_2 is the rate coefficient of the reverse reaction, both being of first order.

As a function of x_A:

$$C_{Ao}\frac{dx_A}{dt} = k_1 C_{Ao}(1 - x_A) - k_2 C_{Ao}\left(M_{RA} + \frac{r}{a}x_A\right)$$

$$\frac{dx_A}{dt} = k_1(1 - x_A) - k_2\left(M_{RA} + \frac{r}{a}x_A\right) \qquad (4.3)$$

Chemical Reaction Kinetics: Concepts, Methods and Case Studies, First Edition.
Jorge Ancheyta.
© 2017 John Wiley & Sons Ltd. Published 2017 by John Wiley & Sons Ltd.

At equilibrium conditions, there is not variation of the conversion with respect to time, that is, the rate of the direct reaction is equal to the rate of the reverse reaction, and also $x_A = x_{Ae}$:

$$\frac{dx_A}{dt} = 0$$

$$k_1(1-x_{Ae}) - k_2\left(M_{RA} + \frac{r}{a}x_{Ae}\right) = 0$$

$$K = \frac{k_1}{k_2} = \frac{M_{RA} + \frac{r}{a}x_{Ae}}{1-x_{Ae}} \tag{4.4}$$

where K is the equilibrium constant.

M_{RA} can be obtained from Eq. (4.4) and then substituting in Eq. (4.3) to incorporate x_{Ae} in the kinetic model:

$$M_{RA} = \frac{ak_1 - x_{Ae}(ak_1 + rk_2)}{ak_2}$$

$$\frac{dx_A}{dt} = k_1(1-x_A) - k_2\left[\frac{ak_1 - x_{Ae}(ak_1 + rk_2)}{ak_2} + \frac{r}{a}x_A\right]$$

$$= k_1(1-x_A) - k_2\left[\frac{ak_1 - x_{Ae}(ak_1 + rk_2) + rk_2x_A}{ak_2}\right]$$

$$\frac{dx_A}{dt} = k_1 - k_1x_A - k_1 + k_1x_{Ae} + \frac{r}{a}k_2x_{Ae} - \frac{r}{a}k_2x_A$$

$$= x_{Ae}\left(k_1 + \frac{r}{a}k_2\right) - x_A\left(k_1 + \frac{r}{a}k_2\right)$$

$$\frac{dx_A}{dt} = (x_{Ae} - x_A)\left(k_1 + \frac{r}{a}k_2\right)$$

Separating variable and integrating:

$$\int_0^{x_A} \frac{dx_A}{x_{Ae} - x_A} = \int_0^t \left(k_1 + \frac{r}{a}k_2\right)dt$$

$$-\ln(x_{Ae} - x_A)\big|_0^{x_A} = \left(k_1 + \frac{r}{a}k_2\right)t$$

$$\ln\left(\frac{x_{Ae}}{x_{Ae} - x_A}\right) = \left(k_1 + \frac{r}{a}k_2\right)t \tag{4.5}$$

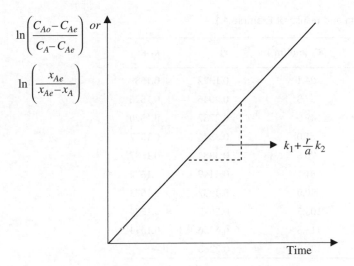

On the y-axis:
$$\ln\left(\frac{C_{Ao}-C_{Ae}}{C_A-C_{Ae}}\right) \quad or$$
$$\ln\left(\frac{x_{Ae}}{x_{Ae}-x_A}\right)$$

Arrow label: $k_1 + \frac{r}{a}k_2$

x-axis: Time

Figure 4.1 Graphical representation of Eqs. (4.5) and (4.6).

And, as a function of C_A:

$$C_{Ae} = C_{Ao}(1-x_{Ae})$$
$$C_A = C_{Ao}(1-x_A)$$
$$\ln\left(\frac{C_{Ao}-C_{Ae}}{C_A-C_{Ae}}\right) = \left(k_1 + \frac{r}{a}k_2\right)t \tag{4.6}$$

If the equilibrium conditions (x_{Ae} or C_{Ae}), and the variation of conversion or concentration with respect to time (x_A or C_A versus t), are known, the term $(k_1 + {}^r/_a k_2)$ can be evaluated for a defined stoichiometry with Eqs. (4.5) or (4.6), as illustrated in Figure 4.1.

On the other hand, with x_{Ae} and Eq. (4.4), the equilibrium constant $K = k_1/k_2$ is calculated. These two relationships give the individual values of k_1 and k_2.

Example 4.1 The reversible reaction of γ-hydroxybutyric acid to form its lactone in aqueous solution can be considered to be first order in both senses, because the concentration of water remains constant. Determine the reaction rate coefficients if, at 25 °C, the experimental results of Table 4.1 were obtained.

$$Acid\ (A) \underset{k_2}{\overset{k_1}{\rightleftarrows}} Water + Lactone\ (R)$$

$$C_{Ao} = 182.3\ mol/m^3$$

Table 4.1 Data and results of Example 4.1.

Time (Ksec)	C_R (mol/m³)	x_A	$k_1 + k_2'$
1.26	24.1	0.1322	0.1589
2.16	37.3	0.2046	0.1526
3.00	49.9	0.2737	0.1570
3.90	61.0	0.3346	0.1577
4.80	70.8	0.3884	0.1587
6.00	81.1	0.4449	0.1572
7.20	90.0	0.4937	0.1573
9.60	103.5	0.5677	0.1574
13.20	115.5	0.6336	0.1544
169.20	132.8	0.7285	–
Average	–	–	0.1567

Solution

The kinetic model is the following:

$$(-r_A) = k_1 C_A - k_2 C_{H_2O} C_R = k_1 C_A - k_2' C_R$$

where

$$k_2' = k_2 C_{H2O}$$

The values of x_A are evaluated with C_R. The last value of time (169.2 Ksec) can be considered as the equilibrium condition.

$$C_R = C_{Ao}\left(M_{RA} + \frac{r}{a}x_A\right) = C_{Ao}x_A$$

Therefore.

$$x_A = \frac{C_R}{C_{Ao}}$$

For this case, Eq. (4.5) is:

$$\ln\left(\frac{x_{Ae}}{x_{Ae} - x_A}\right) = \left(k_1 + k_2'\right)t$$

$$k_1 + k_2' = \frac{1}{t}\ln\left(\frac{x_{Ae}}{x_{Ae} - x_A}\right)$$

The complete results are summarized in Table 4.1. It is observed that the values of the reaction rate coefficients are similar, so that $k_1 + k_2' = 0.1567$.

To evaluate the individual values of k_1 and k_2, it requires the calculation of the equilibrium constant K with Eq. (4.4), and then it is substituted in the reaction rate coefficients previously evaluated:

$$K = \frac{k_1}{k_2'} = \frac{M_{RA} + \dfrac{r}{a}x_{Ae}}{1 - x_{Ae}} = \frac{x_{Ae}}{1 - x_{Ae}} = \frac{0.7285}{1 - 0.7285} = 2.6832$$

$$k_1 + k_2' = k_2'K + k_2' = k_2'(1 + K)$$

$$k_2' = \frac{k_1' + k_2'}{1 + K} = \frac{0.1567}{1 + 2.6832} = 4.254 \times 10^{-2}$$

$$k_1 = k_2'K = (4.254 \times 10^{-2})(2.6832) = 0.1142$$

And, finally, the kinetic model is:

$$(-r_A) = 0.1142 C_A - 4.254 \times 10^{-2} C_R$$

4.2 Reversible Reactions of Second Order

The most common cases for this type of reaction are the following. In all of them, stoichiometric feed composition and pure reactants (no products in the feed) are considered.

$$\text{Case 1}: \quad aA + bB \underset{k_2}{\overset{k_1}{\rightleftharpoons}} rR + sS \quad M_{BA} = b/a \tag{4.7}$$

$$\text{Case 2}: \quad aA \underset{k_2}{\overset{k_1}{\rightleftharpoons}} rR + sS \quad C_{Ro} = C_{So} = 0 \tag{4.8}$$

$$\text{Case 3}: \quad aA \underset{k_2}{\overset{k_1}{\rightleftharpoons}} rR \quad C_{Ro} = 0 \tag{4.9}$$

$$\text{Case 4}: \quad aA + bB \underset{k_2}{\overset{k_1}{\rightleftharpoons}} rR \quad M_{BA} = b/a \tag{4.10}$$

For instance, for Case 1, the kinetic model is:

$$C_A = C_{Ao}(1 - x_A)$$

$$C_B = C_{Ao}\left(M_{BA} - \frac{b}{a}x_A\right) = \frac{b}{a}C_{Ao}(1 - x_A)$$

$$C_R = C_{Ao}\left(M_{RA} + \frac{r}{a}x_A\right) = \frac{r}{a}C_{Ao}x_A$$

$$C_s = C_{Ao}\left(M_{sA} + \frac{s}{a}x_A\right) = \frac{s}{a}C_{Ao}x_A$$

$$(-r_A) = k_1 C_A C_B - k_2 C_R C_S \tag{4.11}$$

where k_1 is the rate coefficient of the direct reaction and k_2 the rate coefficient of the reverse reaction, both of second order.

As a function of x_A:

$$C_{Ao}\frac{dx_A}{dt} = \frac{b}{a}k_1 C_{Ao}^2 (1-x_A)^2 - \left(\frac{rs}{a^2}\right)k_2 C_{Ao}^2 x_A^2$$

$$\frac{dx_A}{dt} = \frac{b}{a}k_1 C_{Ao}(1-x_A)^2 - \left(\frac{rs}{a^2}\right)k_2 C_{Ao}x_A^2 \qquad (4.12)$$

At equilibrium conditions, $dx_A/dt=0$; therefore:

$$K = \frac{k_1}{k_2} = \frac{b\left(\frac{rs}{a^2}\right)x_{Ae}^2}{a(1-x_{Ae})^2} = \frac{rs\,x_{Ae}^2}{ab(1-x_{Ae})^2}$$

$$\frac{dx_A}{dt} = \frac{b}{a}k_1 C_{Ao}(1-x_A)^2 - \left(\frac{rs}{a^2}\right)\frac{k_1}{K}C_{Ao}x_A^2$$

$$\frac{dx_A}{dt} = k_1 C_{Ao}\left[\frac{b}{a}(1-x_A)^2 - \left(\frac{rs}{a^2}\right)\frac{(1-x_{Ae})^2}{\left(\frac{rs}{ab}\right)x_{Ae}^2}x_A^2\right]$$

$$= \frac{b}{a}k_1 C_{Ao}\left\{(1-x_A)^2 - \left[\frac{x_A(1-x_{Ae})}{x_{Ae}}\right]^2\right\} \qquad (4.13)$$

This equation involves the square of a difference: $a^2-b^2 = (a+b)(a-b)$; therefore:

$$\frac{dx_A}{dt} = \frac{b}{a}k_1 C_{Ao}\left\{\left[1-x_A+\frac{x_A(1-x_{Ae})}{x_{Ae}}\right]\left[1-x_A-\frac{x_A(1-x_{Ae})}{x_{Ae}}\right]\right\}$$

$$\frac{dx_A}{dt} = \frac{b\,k_1 C_{Ao}}{a\;x_{Ae}^2}[(x_{Ae}-x_A x_{Ae}+x_A-x_A x_{Ae})(x_{Ae}-x_A x_{Ae}+x_A+x_A x_{Ae})]$$

$$\frac{dx_A}{dt} = \frac{b\,k_1 C_{Ao}}{a\;x_{Ae}^2}\{[x_{Ae}+x_A(1-2x_{Ae})][x_{Ae}-x_A]\} \qquad (4.14)$$

Separating variables:

$$\int_0^{x_A}\frac{dx_A}{[x_{Ae}+x_A(1-2x_{Ae})][x_{Ae}-x_A]} = \frac{k_1 C_{Ao}}{x_{Ae}^2}\int_0^t dt = \frac{b\,k_1 C_{Ao}}{a\;x_{Ae}^2}t \qquad (4.15)$$

The left-hand side of Eq. (4.15) can be written as:

$$\int_0^{x_A} \frac{1}{[x_{Ae}+x_A(1-2x_{Ae})][x_{Ae}-x_A]}dx_A$$

$$=\int_0^{x_A} \frac{A}{[x_{Ae}+x_A(1-2x_{Ae})]}dx_A + \int_0^{x_A} \frac{B}{[x_{Ae}-x_A]}dx_A$$

(4.16)

And then by partial fractions:

$$\frac{A}{[x_{Ae}+x_A(1-2x_{Ae})]}+\frac{B}{[x_{Ae}-x_A]}=\frac{A[x_{Ae}-x_A]+B[x_{Ae}+x_A(1-2x_{Ae})]}{[x_{Ae}+x_A(1-2x_{Ae})][x_{Ae}-x_A]}$$

By comparing the numerators of the right-hand side of this equation with the left-hand side of Eq. (4.16):

$$A[x_{Ae}-x_A]+B[x_{Ae}+x_A(1-2x_{Ae})]=1$$

Which can be rearranged as:

$$[A+B]x_{Ae}+[B-A-2Bx_{Ae}]x_A=1$$

Or:

$$[A+B]x_{Ae}+[B-A-2Bx_{Ae}]x_A=1+[0]x_A$$

By equalling both sides of the equation:

$$[A+B]x_{Ae}=1$$
$$[B-A-2Bx_{Ae}]=0$$

Solving for A and B:

$$A=\frac{1-2x_{Ae}}{2x_{Ae}(1-x_{Ae})}$$

$$B=\frac{1}{2x_{Ae}(1-x_{Ae})}$$

Substituting the values of A and B in Eq. (4.16):

$$\int_0^{x_A}\frac{A}{[x_{Ae}+x_A(1-2x_{Ae})]}dx_A=\left[\frac{1-2x_{Ae}}{2x_{Ae}(1-x_{Ae})}\right]\left(\frac{1}{1-2x_{Ae}}\right)$$

$$\ln[x_{Ae}+x_A(1-2x_{Ae})]|_0^{x_A}=\frac{1}{2x_{Ae}(1-x_{Ae})}\ln\left[\frac{x_{Ae}+x_A(1-2x_{Ae})}{x_{Ae}}\right]$$

$$\int_0^{x_A}\frac{B}{[x_{Ae}-x_A]}dx_A=\frac{1}{2x_{Ae}(1-x_{Ae})}\ln(x_{Ae}-x_A)|_0^{x_A}=-\frac{1}{2x_{Ae}(1-x_{Ae})}\ln\left(\frac{x_{Ae}-x_A}{x_{Ae}}\right)$$

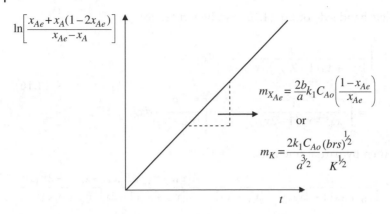

Figure 4.2 Graphic representation of Eqs. (4.17) and (4.19).

Eq. (4.15) is then:

$$\frac{1}{2x_{Ae}(1-x_{Ae})}\ln\left[\frac{x_{Ae}+x_A(1-2x_{Ae})}{x_{Ae}-x_A}\right] = \frac{bk_1C_{Ao}}{a}\frac{t}{x_{Ae}^2}$$

$$\ln\left[\frac{x_{Ae}+x_A(1-2x_{Ae})}{x_{Ae}-x_A}\right] = \frac{2b}{a}k_1C_{Ao}\left(\frac{1-x_{Ae}}{x_{Ae}}\right)t = m_{x_{Ae}}t \qquad (4.17)$$

where $m_{x_{Ae}} = \dfrac{2b}{a}k_1C_{Ao}\left(\dfrac{1-x_{Ae}}{x_{Ae}}\right)$ is the slope of the straight line as a function of x_{Ae}.

This equation can be put as a function of the equilibrium constant K; thus, x_{Ae} can be obtained from Eq. (4.13):

$$x_{Ae} = \frac{abK^{1/2}}{(rs)^{1/2}+(abK)^{1/2}} \qquad (4.18)$$

$$\ln\left[\frac{x_{Ae}+x_A(1-2x_{Ae})}{x_{Ae}-x_A}\right] = \frac{2k_1C_{Ao}}{a^{3/2}}\frac{(brs)^{1/2}}{K^{1/2}}t = m_K t \qquad (4.19)$$

where $m_K = \dfrac{2k_1C_{Ao}}{a^{3/2}}\dfrac{(brs)^{1/2}}{K^{1/2}}$ is the slope of the straight line as a function of K.

Figure 4.2 shows the graphic representation of Eqs. (4.17) and (4.19).

Similarly to Case 1, for Cases 2, 3 and 4 a similar integrated equation is obtained; however, the evaluation of x_{Ae}, K, $m_{x_{Ae}}$ and m_K is different. Table 4.2 shows a summary of the integrated equations for each case.

Example 4.2 The following reversible reaction between sulphuric acid and diethyl sulphate was studied at 22.9 °C. Starting with equal initial

Table 4.2 Integrated equations for reversible reactions of second order.

General equation: $\ln\left[\dfrac{x_{Ae} + x_A(1-2x_{Ae})}{x_{Ae} - x_A}\right] = m_{x_{Ae}}t = m_K t$

Reaction	Equilibrium constant K	Equilibrium conversion x_{Ae}	$m_{x_{Ae}}$	m_K
$aA + bB \underset{k_2}{\overset{k_1}{\rightleftharpoons}} rR + sS$ $M_{BA} = b/a,\ C_{Ro} = C_{So} = 0$	$\dfrac{rs\,x_{Ae}^2}{ab(1-x_{Ae})^2}$	$\dfrac{(abK)^{1/2}}{(rs)^{1/2} + (abK)^{1/2}}$	$\dfrac{2\delta k_1 C_{Ao}(1-x_{Ae})}{a}\,\dfrac{}{x_{Ae}}$	$\dfrac{2k_1 C_{Ao}}{a^{3/2}}\dfrac{(brs)^{1/2}}{K^{1/2}}$
$aA \underset{k_2}{\overset{k_1}{\rightleftharpoons}} rR + sS$ $C_{Ro} = C_{So} = 0$	$\dfrac{rs\,x_{Ae}^2}{a^2(1-x_{Ae})^2}$	$\dfrac{aK^{1/2}}{(rs)^{1/2} + aK^{1/2}}$	$2\bar{k}_1 C_{Ao}\,\dfrac{(1-x_{Ae})}{x_{Ae}}$	$2k_1 C_{Ao}\dfrac{(rs)^{1/2}}{aK^{1/2}}$
$aA \underset{k_2}{\overset{k_1}{\rightleftharpoons}} rR$ $C_{Ro} = 0$	$\dfrac{r^2 x_{Ae}^2}{a^2(1-x_{Ae})^2}$	$\dfrac{aK^{1/2}}{r + aK^{1/2}}$	$2\bar{k}_1 C_{Ao}\,\dfrac{(1-x_{Ae})}{x_{Ae}}$	$2k_1 C_{Ao}\dfrac{r}{aK^{1/2}}$
$aA + bB \underset{k_2}{\overset{k_1}{\rightleftharpoons}} rR$ $M_{BA} = b/a,\ C_{Ro} = 0$	$\dfrac{r^2 x_{Ae}^2}{ab(1-x_{Ae})^2}$	$\dfrac{(abK)^{1/2}}{r + (abK)^{1/2}}$	$\dfrac{2\delta k_1 C_{Ao}(1-x_{Ae})}{a}\,\dfrac{}{x_{Ae}}$	$\dfrac{2k_1 C_{Ao} r\,b^{1/2}}{a^{3/2}}\dfrac{}{K^{1/2}}$

Table 4.3 Data and results of Example 4.2.

Time (sec)	C_R (kmol/m³)	x_A	$m_{xAe} \times 10^4$
1680	0.69	0.0627	0.7151
2880	1.38	0.1255	0.8980
4500	2.24	0.2036	1.0372
5760	2.75	0.2500	1.0702
7620	3.31	0.3009	1.0680
9720	3.81	0.3464	1.0629
10,800	4.11	0.3736	1.1051
12,720	4.45	0.4045	1.1123
16,020	4.86	0.4418	1.1065
19,080	5.15	0.4682	1.1208
22,740	5.35	0.4864	1.1012
24,600	5.42	0.4927	1.0863
11 days	5.80	0.5273	–
Average	–	–	1.0403×10^{-4}

concentrations of reactants (5.5 kmol/m³), the results presented in Table 4.3 were obtained. Find the rate equation for this reaction.

$$H_2SO_4(A) + (C_2H_5)_2SO_4(B) \underset{k_2}{\overset{k_1}{\rightleftharpoons}} 2C_2H_5SO_4H\,(2R)$$

Solution

The reaction can be written as:

$$A + B \underset{k_2}{\overset{k_1}{\rightleftharpoons}} 2R$$

The initial concentration of reactants, the feed molar ratio, and the ratio of stoichiometric coefficients are:

$$C_{Ao} = C_{Bo} = 5.5\,kmol/m^3$$
$$M_{BA} = C_{Bo}/C_{Ao} = 1$$
$$b/a = 1$$

Since

$$M_{BA} = b/a$$

The feed composition is stoichiometric. In addition, $C_{Ro} = 0$.

This reaction is similar to that of Case 4, so that the kinetic model is:

$$(-r_A) = k_1 C_A C_B - k_2 C_R^2$$

The values of x_A are determined with the values of C_R. The last data, at time of 11 days, can be considered as the equilibrium. Therefore:

$$C_R = C_{Ao}\left(M_{RA} + \frac{r}{a}x_A\right) = 2C_{Ao}x_A$$

$$x_A = \frac{C_R}{2C_{Ao}}$$

The corresponding integrated equation is Eq. (4.17):

$$\ln\left[\frac{x_{Ae} + x_A(1-2x_{Ae})}{x_{Ae} - x_A}\right] = \frac{2b}{a}k_1 C_{Ao}\left(\frac{1-x_{Ae}}{x_{Ae}}\right)t$$

$$m_{x_{Ae}} = \frac{1}{t}\ln\left[\frac{x_{Ae} + x_A(1-2x_{Ae})}{x_{Ae} - x_A}\right]$$

The complete results are summarized in Table 4.3. It is observed that the values of $m_{x_{Ae}}$ are similar without any tendency, so that the average value is 1.0403×10^{-4}. From this value and with the definition of $m_{x_{Ae}}$ for the studied reaction (Table 4.2), k_1 can be calculated as follows:

$$m_{x_{Ae}} = \frac{2b}{a}k_1 C_{Ao}\left(\frac{1-x_{Ae}}{x_{Ae}}\right) = 1.0403 \times 10^{-4}$$

$$k_1 = \frac{1.0403 \times 10^{-4}}{\frac{2b}{a}C_{Ao}\left(\frac{1-x_{Ae}}{x_{Ae}}\right)} = \frac{1.0403 \times 10^{-4}}{2(5.5)\left(\frac{1-0.5273}{0.5273}\right)} = 1.0549 \times 10^{-5}$$

The value of k_2 can be determined from the value of K (Table 4.2):

$$K = \frac{k_1}{k_2} = \frac{r^2 x_{Ae}^2}{ab(1-x_{Ae})^2} = \frac{2^2(0.5273)^2}{(1)(1)(1-0.5273x_{Ae})^2} = 4.9774$$

$$k_2 = \frac{k_1}{K} = \frac{1.0549 \times 10^{-5}}{4.9774} = 2.1194 \times 10^{-6}$$

And, finally, the kinetic model is:

$$(-r_A) = 1.0549 \times 10^{-5} C_A C_B - 2.1194 \times 10^{-6} C_R^2$$

4.3 Reversible Reactions with Combined Orders

When the orders of the direct and reverse reactions are different, it is a reversible reaction with combined orders. The most common cases are the reversible reactions that exhibit the following stoichiometry and kinetic models:

$$\text{Case 1}: aA \underset{k_2}{\overset{k_1}{\rightleftharpoons}} rR \quad (-r_A) = k_1 C_A - k_2 C_R^2 \tag{4.20}$$

$$\text{Case 2}: aA \underset{k_2}{\overset{k_1}{\rightleftharpoons}} rR \quad (-r_A) = k_1 C_A^2 - k_2 C_R \tag{4.21}$$

$$\text{Case 3}: aA + bB \underset{k_2}{\overset{k_1}{\rightleftharpoons}} rR \quad (-r_A) = k_1 C_A C_B - k_2 C_R \tag{4.22}$$

$$\text{Case 4}: aA \underset{k_2}{\overset{k_1}{\rightleftharpoons}} rR + sS \quad (-r_A) = k_1 C_A - k_2 C_R C_S \tag{4.23}$$

For instance, for Case 1, the kinetic model given by Eq. (4.20) as a function of x_A is:

$$C_A = C_{Ao}(1 - x_A)$$

$$C_R = C_{Ao}\left(M_{RA} + \frac{r}{a}x_A\right)$$

$$C_{Ao}\frac{dx_A}{dt} = k_1 C_{Ao}(1 - x_A) - k_2 C_{Ao}^2\left(M_{RA} + \frac{r}{a}x_A\right)^2$$

$$\frac{dx_A}{dt} = k_1(1 - x_A) - k_2 C_{Ao}\left(M_{RA} + \frac{r}{a}x_A\right)^2$$

where k_1 is the rate coefficient of the first-order direct reaction, and k_2 is the rate coefficient of the second-order reverse reaction.

Including the equilibrium constant $K = k_1/k_2$ in the previous equation and arranging the terms:

$$\frac{dx_A}{dt} = k_1(1 - x_A) - \frac{k_1}{K}C_{Ao}\left(M_{RA} + \frac{r}{a}x_A\right)^2$$

$$= k_1\left(1 - x_A\frac{-M_{RA}^2 C_{Ao}}{K} - \frac{2rM_{RA}C_{Ao}}{aK}x_A - \frac{r^2 C_{Ao}}{a^2 K}x_A^2\right)$$

$$\frac{dx_A}{dt} = k_1\left[\left(-\frac{r^2 C_{Ao}}{a^2 K}\right)x_A^2 + \left(-1 - \frac{2rM_{RA}C_{Ao}}{aK}\right)x_A + \left(\frac{1 - M_{RA}^2 C_{Ao}}{K}\right)\right]$$

$$= k_1\left(ax_A^2 + bx_A + c\right)$$

$$\tag{4.24}$$

where

$$a = -\frac{r^2 C_{Ao}}{a^2 K}$$

$$b = -1 - \frac{2r M_{RA} C_{Ao}}{aK}$$

$$c = 1 - \frac{M_{RA}^2 C_{Ao}}{K}$$

Separating variables in Eq. (4.24) gives:

$$\int_0^{x_A} \frac{dx_A}{a\, x_A^2 + b x_A + c} = k_1 \int_0^t dt = k_1 t$$

$$k_1 = \frac{1}{t} \int_0^{x_A} \frac{dx_A}{a\, x_A^2 + b x_A + c} \tag{4.25}$$

The denominator of Eq. (4.25) can be written as:

$$a x_A^2 + b x_A + c = (x_A - x_{A1})(x_A - x_{A2})$$

The solution of this equation is:

$$x_{Ai} = \frac{-b \pm \sqrt{b^2 - 4ac}}{2a}$$

where $x_{Ai} = x_{A1}, x_{A2}$ are the roots of the quadratic equation.

Therefore, three solutions are possible depending on the value of b^2 with respect to $4ac$:

1) $b^2 > 4ac$
2) $b^2 < 4ac$
3) $b^2 = 4ac$

Table 4.4 shows the particular solutions for each of the cases presented previously as functions of constants a, b and c.

The previous approach and equations can be also applied for some reversible reactions of second order, different to those reported in Section 4.2, as they are considered as non-stoichiometric feed composition with concentrations of products different to zero. The particular solutions of the following cases are presented in Table 4.5.

$$\text{Case 5}: aA + bB \underset{k_2}{\overset{k_1}{\rightleftarrows}} rR + sS \quad M_{BA} \neq b/a, C_{Ro} \neq 0, C_{So} \neq 0$$

Table 4.4 Integrated equations for reversible reactions with combined orders.

$b^2 > 4ac$: $\quad k_1 = \dfrac{1}{D_1 t}\ln\left[\dfrac{(2ax_A + b - D_1)(b + D_1)}{(2ax_A + b + D_1)(b - D_1)}\right]$ $\qquad$ where $D_1 = \sqrt{b^2 - 4ac}$

$b^2 < 4ac$: $\quad k_1 = \dfrac{2}{D_2 t}\left[\tan^{-1}\left(\dfrac{2ax_A + b}{D_2}\right) - \tan^{-1}\left(\dfrac{b}{D_2}\right)\right]$ $\qquad$ where $D_2 = \sqrt{4ac - b^2}$

$b^2 = 4ac$: $\quad k_1 = \dfrac{1}{t}\left[\dfrac{4ax_A}{b(2ax_A + b)}\right]$

Reaction	Kinetic model	K	a	b	c
$aA \underset{k_2}{\overset{k_1}{\rightleftharpoons}} rR$	$(-r_A) = k_1 C_A - k_2 C_R^2$	$\dfrac{C_{Ao}\left(M_{RA} + \frac{r}{a}x_{Ae}\right)}{1 - x_{Ae}}$	$-\dfrac{r^2 C_{Ao}}{a^2 K}$	$-1 - \dfrac{2r M_{RA} C_{Ao}}{aK}$	$1 - \dfrac{M_{RA}^2 C_{Ao}}{K}$
$aA \underset{k_2}{\overset{k_1}{\rightleftharpoons}} rR$	$(-r_A) = k_1 C_A^2 - k_2 C_R$	$\dfrac{M_{RA} + \frac{r}{a}x_{Ae}}{C_{Ao}(1 - x_{Ae})^2}$	C_{Ao}	$-2C_{Ao} - \dfrac{r}{aK}$	$C_{Ao} - \dfrac{M_{RA}}{K}$
$aA + bB \underset{k_2}{\overset{k_1}{\rightleftharpoons}} rR$	$(-r_A) = k_1 C_A C_B - k_2 C_R$	$\dfrac{M_{RA} + \frac{r}{a}x_{Ae}}{C_{Ao}(1 - x_{Ae})\left(M_{BA}\frac{b}{a} - x_{Ae}\right)}$	$\dfrac{b}{a} C_{Ao}$	$-C_{Ao}\left(\dfrac{b}{a} + M_{RA}\right) - \dfrac{r}{aK}$	$C_{Ao}M_{BA} - \dfrac{M_{RA}}{K}$
$aA \underset{k_2}{\overset{k_1}{\rightleftharpoons}} rR + sS$	$(-r_A) = k_1 C_A - k_2 C_R C_S$	$\dfrac{C_{Ao}\left(M_{RA} + \frac{r}{a}x_{Ae}\right)\left(M_{SA} + \frac{s}{a}x_{Ae}\right)}{1 - x_{Ae}}$	$-\dfrac{rs C_{Ao}}{a^2 K}$	$-1 - \dfrac{C_{Ao}}{aK}(sM_{RA} - rM_{SA})$	$1 - \dfrac{M_{RA}M_{SA}C_{Ao}}{K}$

Table 4.5 Integrated equations for reversible reactions of second order with non-stoichiometric feed composition.

$b^2 > 4ac: \quad k_1 = \dfrac{1}{D_1 t}\ln\left[\dfrac{(2ax_A + b - D_1)(b + D_1)}{(2ax_A + b + D_1)(b - D_1)}\right] \qquad \text{where } D_1 = \sqrt{b^2 - 4ac}$

$b^2 < 4ac: \quad k_1 = \dfrac{2}{D_2 t}\left[\tan^{-1}\left(\dfrac{2ax_A + b}{D_2}\right) - \tan^{-1}\left(\dfrac{b}{D_2}\right)\right] \quad \text{where } D_2 = \sqrt{4ac - b^2}$

$b^2 = 4ac: \quad k_1 = \dfrac{1}{t}\left[\dfrac{4ax_A}{b(2ax_A + b)}\right]$

Reaction	Kinetic model	K	a	b	c
$aA + bB \underset{k_2}{\overset{k_1}{\rightleftharpoons}} R + S$	$(-r_A) = k_1 C_A C_B - k_2 C_R C_S$ $M_{BA} \neq b/a,\ C_{Ro} \neq 0,\ C_{So} \neq 0$	$\dfrac{\left(M_{RA} + \frac{r}{a}x_{Ae}\right)\left(M_{SA} + \frac{s}{a}x_{Ae}\right)}{(1 - x_{Ae})\left(M_{BA} - \frac{b}{a}x_{Ae}\right)}$	$C_{Ao}\left(\dfrac{b}{a} + \dfrac{rs}{a^2 K}\right)$	$-C_{Ao}\left(\dfrac{b}{a} + M_{BA} + \dfrac{sM_{RA} + rM_{BS}}{aK}\right)$	$C_{Ao}\left(M_{BA} - \dfrac{M_{RA}M_{SA}}{K}\right)$
$aA \underset{k_2}{\overset{k_1}{\rightleftharpoons}} rR + sS$	$(-r_A) = k_1 C_A^2 - k_2 C_R C_S$ $C_{Ro} \neq 0,\ C_{So} \neq 0$	$\dfrac{\left(M_{RA} + \frac{r}{a}x_{Ae}\right)\left(M_{SA} + \frac{s}{a}x_{Ae}\right)}{(1 - x_{Ae})^2}$	$C_{Ao}\left(1 + \dfrac{rs}{a^2 K}\right)$	$-C_{Ao}\left(2 + \dfrac{sM_{RA} + rM_{BS}}{aK}\right)$	$C_{Ao}\left(1 - \dfrac{M_{RA}M_{SA}}{K}\right)$
$aA \underset{k_2}{\overset{k_1}{\rightleftharpoons}} rR$	$(-r_A) = k_1 C_A^2 - k_2 C_R^2$ $C_{Ro} \neq 0$	$\dfrac{\left(M_{RA} + \frac{r}{a}x_{Ae}\right)^2}{(1 - x_{Ae})^2}$	$C_{Ao}\left(1 + \dfrac{r^2}{a^2 K}\right)$	$-2C_{Ao}\left(1 + \dfrac{rM_{RA}}{aK}\right)$	$C_{Ao}\left(\dfrac{1 - M_{RA}^2}{K}\right)$
$aA + bB \underset{k_2}{\overset{k_1}{\rightleftharpoons}} rR$	$(-r_A) = k_1 C_A C_B - k_2 C_R^2$ $M_{BA} \neq b/a,\ C_{Ro} \neq 0$	$\dfrac{\left(M_{RA} + \frac{r}{a}x_{Ae}\right)^2}{(1 - x_{Ae})\left(M_{BA}\frac{b}{a} - x_{Ae}\right)}$	$C_{Ao}\left(\dfrac{b}{a} + \dfrac{r^2}{a^2 K}\right)$	$-C_{Ao}\left(\dfrac{b}{a} + M_{BA} + \dfrac{2rM_{RA}}{aK}\right)$	$C_{Ao}\left(\dfrac{M_{BA} - M_{RA}^2}{K}\right)$

$$\text{Case 6}: aA \underset{k_2}{\overset{k_1}{\rightleftharpoons}} rR + sS \quad C_{Ro} \neq C_{So} \neq 0$$

$$\text{Case 7}: aA \underset{k_2}{\overset{k_1}{\rightleftharpoons}} rR \quad C_{Ro} \neq 0$$

$$\text{Case 8}: aA + bB \underset{k_2}{\overset{k_1}{\rightleftharpoons}} rR \quad M_{BA} \neq b/a, C_{Ro} \neq 0$$

Example 4.3   In the following reversible reaction carried out in liquid phase at 25°C, the data reported in Table 4.6 were obtained starting with pure reactant A with an initial concentration of 0.05 mol/lt. Find the reaction rate equation.

$$A \underset{k_2}{\overset{k_1}{\rightleftharpoons}} 2R$$

Solution

Assume that the reaction follows a kinetics such as in Case 1. First, the equilibrium constant K and the molar feed ratio M_{RA} are required.

$$C_{Ao} = 0.05 \, mol/lt$$

$$C_{Ro} = 0$$

$$M_{RA} = \frac{C_{Ro}}{C_{Ao}} = \frac{0}{0.05} = 0$$

From Table 4.6, it is observed that $x_{Ae} = 0.82$. K is obtained from Table 4.4:

Table 4.6 Data and results of Example 4.3.

Time (min)	x_A	$k_1 \times 10^2$ Case 1
5	0.28	6.635
10	0.46	6.389
15	0.58	6.247
20	0.66	6.148
25	0.72	6.257
∞	0.82	–
Average	–	6.335×10^{-2}

$$K = \frac{C_{Ao}\left(M_{RA} + \frac{r}{a}x_{Ae}\right)^2}{1 - x_{Ae}} = \frac{4C_{Ao}\,x_{Ae}^2}{1 - x_{Ae}} = \frac{4(0.05)(0.82)^2}{1 - 0.82} = 0.7471$$

For Case 1 of Table 4.4:

$$a = -\frac{r^2 C_{Ao}}{a^2 K} = -\frac{2^2(0.05)}{1^2(0.7471)} = -0.2677$$

$$b = -1 - \frac{2rM_{RA}C_{Ao}}{aK} = -1$$

$$c = 1 - \frac{M_{RA}^2 C_{Ao}}{K} = 1$$

Therefore:

$$b^2 = (-1)^2 = 1$$
$$4ac = 4(-0.2677)(1) = -1.0708$$

Then, for this case:

$$b^2 > 4ac$$

Thus, the value of k_1 is evaluated with the following equation:

$$k_1 = \frac{1}{D_1 t}\ln\left[\frac{(2ax_A + b - D_1)(b + D_1)}{(2ax_A + b + D_1)(b - D_1)}\right]$$

Table 4.6 summarized the calculated values of k_1 using the previous equation. It is observed that the values do not show any tendency and they are close, having an average value of k_1 of 6.335×10^{-2}.

The value of k_2 is evaluated with the equilibrium constant as follows:

$$K = \frac{k_1}{k_2} = 0.7471$$

$$k_2 = \frac{k_1}{K} = \frac{6.335 \times 10^{-2}}{0.7471} = 8.480 \times 10^{-2}$$

And, finally, the kinetic model is:

$$(-r_A) = 6.335 \times 10^{-2}C_A - 8.480 \times 10^{-2}C_R^2$$

5

Complex Reactions

Most of the chemical reactions of commercial interest are complex; that is, they follow mechanisms that involve various elemental steps, and therefore they follow complex kinetics.

If the mechanisms involve one or more reactants that disappear by means of one or more reactions, they are called *simultaneous* or *parallel reactions*. On the contrary, if during the disappearance of one or more reactants one product is produced which at the same time disappears to form another and so on, the reaction is called *in series* or *consecutive*.

There are also reactions that involve various combinations of in-series or parallel reactions that can be called *in-series–parallel reactions*.

5.1   Yield and Selectivity

Differently from irreversible or reversible reactions that take place with only one reaction path, in which the conversion is the most important variable to measure, in complex reactions the most important variables are product yield and selectivity, because one product can be obtained by diverse reaction paths.

Product yield and *selectivity* are defined as:

$$Yield = Y_i = \frac{Moles\ of\ component\ i\ at\ any\ time}{Initial\ moles\ of\ the\ limiting\ reactant} \tag{5.1}$$

$$Selectivity = S_{ij} = \frac{Moles\ of\ component\ i\ at\ any\ time}{Transformed\ moles\ of\ component\ j} \tag{5.2}$$

If, in a complex reaction, A is the limiting reactant, the yield of the product R as a function of concentrations is:

$$Y_R = \frac{C_R}{C_{Ao}} \tag{5.3}$$

Chemical Reaction Kinetics: Concepts, Methods and Case Studies, First Edition.
Jorge Ancheyta.
© 2017 John Wiley & Sons Ltd. Published 2017 by John Wiley & Sons Ltd.

The selectivity of the product R with respect to the reactant A is:

$$S_{RA} = \frac{C_R}{C_{Ao}x_A} = \frac{C_R}{C_{Ao} - C_A} \tag{5.4}$$

The relationship between product selectivity and yield with respect to the same component is obtained by dividing Eqs. (5.4) and (5.3):

$$S_{RA} = \frac{Y_R}{x_A} \tag{5.5}$$

In addition, since the product yields are referred to the same limiting reactant, the sum of all yields must be equal to unity:

$$\sum_{i=1}^{n} Y_i = 1 \tag{5.6}$$

It is common to represent Eq. (5.5) in a graphical form with x_A versus Y_R (conversion versus product yield), in which the slope of the resulting curve is the product selectivity.

For instance, for the following in series reaction, Figure 5.1 shows the curves of conversion of A versus yield of R for different values of the ratio of reaction rate coefficients k_1 and k_2.

$$A \xrightarrow{k_1} R \xrightarrow{k_2} S$$

$$K = \frac{k_1}{k_2}$$

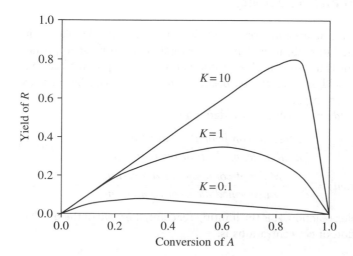

Figure 5.1 Profiles of product yield and conversion for a reaction in series with different values of K.

5.2   Simultaneous or Parallel Irreversible Reactions

5.2.1   Simultaneous Reactions with the Same Order

5.2.1.1   Case 1: Reactions with Only One Reactant

The simplest reaction that occurs in simultaneous mechanisms is the transformation of the reactant A to give different products following a kinetics of order n:

$$aA \xrightarrow{k_1} rR + ...$$
$$aA \xrightarrow{k_2} sS + ...$$
$$aA \xrightarrow{k_n} zZ + ...$$

or

$$aA \begin{array}{c} \nearrow rR+... \\ \xrightarrow{k_1} \\ \xrightarrow{k_2} sS+... \\ \xrightarrow{k_n} \\ \searrow zZ+... \end{array}$$

The reaction rate for each component is:

$$(-r_A) = -\frac{dC_A}{dt} = k_1 C_A^n + k_2 C_A^n + ... + k_n C_A^n =$$
$$(k_1 + k_2 + ... + k_n)C_A^n = k_0 C_A^n \tag{5.7}$$

$$(r_R) = \frac{dC_R}{dt} = k_1 C_A^n \tag{5.8}$$

$$(r_S) = \frac{dC_S}{dt} = k_2 C_A^n \tag{5.9}$$

$$(r_Z) = \frac{dC_Z}{dt} = k_n C_A^n \tag{5.10}$$

Figure 5.2 shows the typical profiles of concentrations with respect to time for the previous reaction.

Eq. (5.7) is used to evaluate k_0, which is similar to the case of irreversible reactions with one component discussed in Chapter 2. Therefore, the resulting integrated equation is:

$$\ln \frac{C_{Ao}}{C_A} = k_0 t \quad \text{For } n = 1 \tag{5.11}$$

$$C_A^{1-n} - C_{Ao}^{1-n} = (n-1)k_0 t \quad \text{For } n \neq 1 \tag{5.12}$$

Figure 5.3 illustrates the calculation of k_0 depending on the reaction order using Eqs. (5.11) and (5.12).

The evaluation of the other reaction rate coefficients requires the utilization of the reaction rate expressions for the other components.

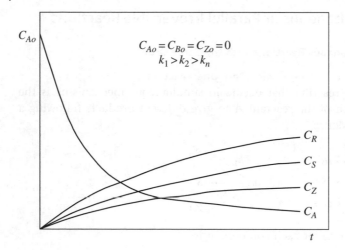

Figure 5.2 Typical profile of concentration for simultaneous or parallel irreversible reactions.

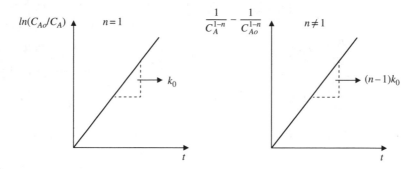

Figure 5.3 Calculation of k_0 for simultaneous irreversible reactions.

Dividing Eqs. (5.9) and (5.8), and integrating, the following expressions are obtained to evaluate the ratio $R_{21} = k_2/k_1$:

$$\frac{dC_S/dt}{dC_R/dt} = \frac{dC_S}{dC_R} = \frac{k_2\, C_A^n}{k_1\, C_A^n} = \frac{k_2}{k_1}$$

$$\int_{C_{So}}^{C_S} dC_S = \frac{k_2}{k_1} \int_{C_{Ro}}^{C_R} dC_R$$

$$C_S - C_{So} = \frac{k_2}{k_1}(C_R - C_{Ro}) = R_{21}(C_R - C_{Ro})$$

$$R_{21} = \frac{k_2}{k_1} = \frac{C_S - C_{So}}{C_R - C_{Ro}} \tag{5.13}$$

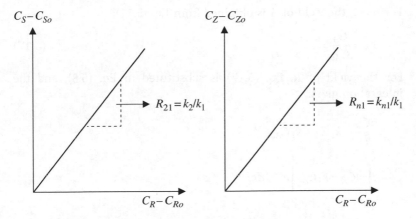

Figure 5.4 Calculation of R_{21} and R_{n1} for simultaneous irreversible reactions.

Similarly, dividing Eqs. (5.10) and (5.8), and integrating, the ratio $R_{n1} = k_n/k_1$ is obtained:

$$R_{n1} = \frac{k_n}{k_1} = \frac{C_Z - C_{Zo}}{C_R - C_{Ro}} \tag{5.14}$$

From Eqs. (5.13) and (5.14), the ratios R_{21} and R_{n1} are calculated as illustrated in Figure 5.4.

Substituting the values of the ratios $R_{21} = k_2/k_1$ and $R_{n1} = k_n/k_1$ in k_0 gives:

$$k_0 = k_1 + k_2 + \ldots + k_n = k_1 + k_1 R_{21} + \ldots + k_1 R_{n1}$$

$$k_0 = (1 + R_{21} + \ldots + R_{n1})k_1$$

And, finally, k_1 is evaluated with:

$$k_1 = \frac{k_0}{1 + R_{21} + \ldots + R_{n1}} = \frac{k_0}{1 + \sum_{j=2}^{n} R_{j1}} = \frac{k_0}{1 + \sum_{j=2}^{n} \frac{k_j}{k_1}} \tag{5.15}$$

In this way, if k_1 is known, the other reaction rate coefficients (k_2, ..., k_n) are obtained with the ratios of R_{21}, ..., R_{n1} in the following manner:

$$k_j = R_{j1}k_1 \quad \text{For } j = 2, n \tag{5.16}$$

5.2.1.1.1 *Product Yields*
The yield of A is calculated with Eqs. (5.11) and (5.12) depending on the reaction order.

- For $n = 1$, the yield of A is obtained from Eq. (5.11):

$$Y_A = \frac{C_A}{C_{Ao}} = e^{-k_0 t} \tag{5.17}$$

For the yield of R, Eq. (5.17) is substituted in Eq. (5.8), and the integration gives:

$$\frac{dC_R}{dt} = k_1 C_A = k_1 C_{Ao} e^{-k_0 t}$$

$$\int_{C_{Ro}}^{C_R} dC_R = k_1 C_{Ao} \int_0^t e^{-k_0 t} dt$$

$$C_R - C_{Ro} = -\frac{k_1}{k_0} C_{Ao} \left(e^{-k_0 t} - 1 \right)$$

$$Y_R = \frac{C_R}{C_{Ao}} = \frac{C_{Ro}}{C_{Ao}} + \frac{k_1}{k_0} \left(1 - e^{-k_0 t} \right) \tag{5.18}$$

Similarly, for the yields of R and Z:

$$Y_S = \frac{C_S}{C_{Ao}} = \frac{C_{So}}{C_{Ao}} + \frac{k_2}{k_0} \left(1 - e^{-k_0 t} \right) \tag{5.19}$$

$$Y_Z = \frac{C_Z}{C_{Ao}} = \frac{C_{Zo}}{C_{Ao}} + \frac{k_n}{k_0} \left(1 - e^{-k_0 t} \right) \tag{5.20}$$

- For $n = 2$, the yield of A is obtained from Eq. (5.12):

$$\frac{1}{C_A} - \frac{1}{C_{Ao}} = k_0 t$$

$$Y_A = \frac{C_A}{C_{Ao}} = \frac{1}{1 + C_{Ao} k_0 t} \tag{5.21}$$

For the yield of R, Eq. (5.21) is substituted in Eq. (5.8):

$$\frac{dC_R}{dt} = k_1 C_A^2 = k_1 \left(\frac{C_{Ao}}{1 + C_{Ao} k_0 t} \right)^2$$

$$\int_{C_{ro}}^{C_R} dC_R = k_1 C_{Ao}^2 \int_0^t \frac{1}{(1 + C_{Ao} k_0 t)^2} dt$$

$$C_R - C_{Ro} = k_1 \left(\frac{C_{Ao} t}{1 + C_{Ao} k_0 t} \right)$$

And, finally:

$$Y_R = \frac{C_R}{C_{Ao}} = \frac{C_{Ro}}{C_{Ao}} + \frac{k_1 t}{1 + C_{Ao}k_0 t} \tag{5.22}$$

In the same manner, for the yields of S and Z:

$$Y_S = \frac{C_S}{C_{Ao}} = \frac{C_{So}}{C_{Ao}} + \frac{k_2 t}{1 + C_{Ao}k_0 t} \tag{5.23}$$

$$Y_Z = \frac{C_Z}{C_{Ao}} = \frac{C_{Zo}}{C_{Ao}} + \frac{k_n t}{1 + C_{Ao}k_0 t} \tag{5.24}$$

Example 5.1 The following reaction in parallel was studied at 204.5 °C. The experimental data of weight percentage of each component at different reaction times are presented in Table 5.1. Find the kinetic model.

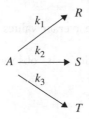

Table 5.1 Data and results of Example 5.1.

Time (min)	y_A (wt%)	y_R (wt%)	y_S (wt%)	y_T (wt%)	$k_0 \times 10^4$	R_{21}	R_{31}
0	1.000	0.000	0.000	0.000	–	–	–
440	0.859	0.082	0.051	0.008	3.454	0.6219	0.0976
825	0.743	0.156	0.092	0.009	3.601	0.5897	0.0577
1200	0.651	0.215	0.121	0.013	3.577	0.5628	0.0605
1500	0.586	0.255	0.146	0.013	3.563	0.5725	0.0510
2040	0.481	0.319	0.183	0.017	3.588	0.5737	0.0533
3060	0.321	0.420	0.237	0.022	3.713	0.5643	0.0524
6060	0.112	0.547	0.315	0.026	3.613	0.5759	0.0475
Average	–	–	–	–	3.587×10^{-4}	0.5801	0.0600

Solution

Assuming that all the reactions occur with the same order, for instance for $n = 1$, the reaction rate coefficient is evaluated with Eq. (5.11) by changing concentrations by weight fractions. For the first data:

$$k_0 = \frac{1}{t}\ln\frac{y_{Ao}}{y_A} = \frac{1}{440}\ln\frac{1}{0.859} = 3.454 \times 10^{-4}$$

The other calculated values of k_0 are summarized in Table 5.1. It can be noted that these values are similar with an average value of 3.587×10^{-4}. This indicates that all the reactions follow first-order kinetics.

To calculate k_1, the ratios of R_{21} and R_{31} need to be determined with Eqs. (5.13) and (5.14) as follows:

$$R_{21} = \frac{y_S - y_{So}}{y_R - y_{Ro}} = \frac{0.051 - 0}{0.082 - 0} = 0.6219$$

$$R_{31} = \frac{y_T - y_{To}}{y_R - y_{Ro}} = \frac{0.008 - 0}{0.082 - 0} = 0.0976$$

The complete results are reported in Table 5.1. The average values of R_{21} and R_{31} are 0.5801 and 0.0600, respectively.

A graphical representation of these calculations is observed in Figure 5.5.

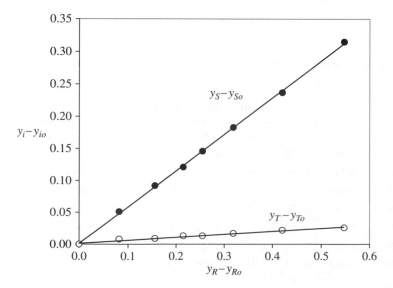

Figure 5.5 Calculation of R_{21} and R_{31}.

The value of k_1 is evaluated with Eq. (5.15):

$$k_1 = \frac{k_0}{1 + R_{21} + R_{31}} = \frac{3.587 \times 10^{-4}}{1 + 0.5801 + 0.0600} = 2.187 \times 10^{-4}$$

The other reaction rate coefficients are calculated with Eq. (5.16):

$$k_2 = R_{21}k_1 = (0.5801)(2.187 \times 10^{-4}) = 1.269 \times 10^{-4}$$
$$k_3 = R_{31}k_1 = (0.0600)(2.187 \times 10^{-4}) = 1.312 \times 10^{-5}$$

And, finally, the kinetic model is:

$$(-r_A) = 3.434 \times 10^{-4}C_A$$

5.2.1.2 Case 2: Reactions with Two Reactants

$$aA + bB \xrightarrow{k_1} rR + \dots$$
$$aA + bB \xrightarrow{k_2} sS + \dots \quad \text{or} \quad aA + bB \begin{array}{c} \overset{k_1}{\nearrow} rR + \dots \\ \overset{k_2}{\longrightarrow} sS + \dots \\ \underset{k_n}{\searrow} zZ + \dots \end{array}$$
$$aA + bB \xrightarrow{k_n} zZ + \dots$$

The reaction rate for each component is:

$$-\frac{dC_A}{dt} = k_1 C_A^\alpha C_B^\beta + k_2 C_A^\alpha C_B^\beta + \dots + k_n C_A^\alpha C_B^\beta =$$
$$(k_1 + k_2 + \dots + k_n)C_A^\alpha C_B^\beta = k_0 C_A^\alpha C_B^\beta \tag{5.25}$$

$$\frac{dC_R}{dt} = k_1 C_A^\alpha C_B^\beta \tag{5.26}$$

$$\frac{dC_S}{dt} = k_2 C_A^\alpha C_B^\beta \tag{5.27}$$

$$\frac{dC_Z}{dt} = k_n C_A^\alpha C_B^\beta \tag{5.28}$$

To evaluate k_0, Eq. (5.25) is used, which is similar to the case of irreversible reactions with two components discussed in Chapter 3. Therefore, the integration will depend on the type of feed composition (stoichiometric, non-stoichiometric or with a reactant in excess), or on whether the reaction is (or is not) elemental, according to the equations reported in Tables 3.2 and 3.3.

The evaluation of the other reaction rate coefficients is done with Eqs. (5.26)–(5.28). By dividing Eqs. (5.27) and (5.28) by Eq. (5.26) and integrating them, the following expressions are obtained for calculating the ratios of $R_{j1} = k_j/k_1$, similarly to Case 1:

$$\frac{dC_S \big/ dt}{dC_R \big/ dt} = \frac{dC_S}{dC_R} = \frac{k_2 \, C_A^\alpha \, C_A^\beta}{k_1 \, C_A^\alpha \, C_A^\beta} = \frac{k_2}{k_1}$$

$$\int_{C_{So}}^{C_S} dC_S = \frac{k_2}{k_1} \int_{C_{Ro}}^{C_R} dC_R$$

$$C_S - C_{So} = \frac{k_2}{k_1}(C_R - C_{Ro}) = R_{21}(C_R - C_{Ro})$$

In the same manner:

$$C_Z - C_{Zo} = \frac{k_n}{k_1}(C_R - C_{Ro}) = R_{n1}(C_R - C_{Ro})$$

Thus, the ratios of $R_{j1} = k_j/k_1$ are determined with Eqs. (5.13) and (5.14), similarly to Case 1, as illustrated in Figure 5.4.

Finally, the values of k_1 and k_j are calculated with Eqs. (5.15) and (5.16), respectively.

Example 5.2   The following reaction in liquid phase was studied at 51 °C. Using equimolar feed composition of reactants A and B, the experimental data reported in Table 5.2 were obtained. Find the kinetic model.

$$A + B \xrightarrow{k_1} R$$
$$A + B \xrightarrow{k_2} S$$

Table 5.2 Data and results of Example 5.2.

Time (min)	C_A (kmol/m³)	$k' \times 10^3$
0	0.181	–
270	0.141	5.805
360	0.131	5.858
540	0.119	5.331
630	0.111	5.530
1440	0.0683	6.331
1650	0.0644	6.062
1800	0.0603	6.144
Average	–	5.866×10^{-3}

Solution

Since the feed composition is equimolar and the stoichiometric coefficients of reactants A and B are the same, the feed composition is stoichiometric. Moreover, because the reaction is carried out in liquid phase, the reaction is conducted at constant density. Assuming that the two reactions follow a second order, to calculate k_0, the equations reported in Table 3.2 are used. For $n \neq 1$:

$$k' = \frac{1}{(n-1)t}\left[C_A^{1-n} - C_{Ao}^{1-n}\right]$$

$$k' = \left(\frac{b}{a}\right)^b k_0 \qquad \alpha = a, \beta = b, n = a + b \qquad \text{Elemental reaction}$$

$$k' = \left(\frac{b}{a}\right)^\beta k_0 \qquad \alpha \neq a, \beta \neq b, n = \alpha + \beta \qquad \text{Non-elemental reaction}$$

Assuming that the reaction follows a second global order ($n = 2$) for the first data:

$$k' = \frac{1}{t}\left[C_A^{-1} - C_{Ao}^{-1}\right] = \frac{1}{270}\left(\frac{1}{0.141} - \frac{1}{0.181}\right) = 5.805 \times 10^{-3}$$

Table 5.2 reports the complete results. Constant values of k' are observed with an average value of 5.866×10^{-3}, so that the assumed reaction order is correct. As the ratio of stoichiometric coefficients of the reactants b/a is equal to one, the calculated value of k' is the same as k_0 according to the previous equation, no matter if the reaction is elemental or not.

The individual reaction orders and reaction rate coefficients cannot be determined because the concentrations of R and S are required.

For the two cases of irreversible simultaneous reactions with the same order, with one or two reactants, as presented in this chapter, it is possible to determine the ratio between reaction rate coefficients ($R_{j1} = k_j/k_1$) without knowing the order of reactions, but only if the order is the same for all the reactions.

5.2.2 Simultaneous Reactions with Combined Orders

The most common case is when one of the reactions occurs with an order different from the others. For instance, if in the following simultaneous reaction R is produced with second-order kinetics, and S follows first-order kinetics, the reaction rate equations are:

$$aA \begin{array}{c} \nearrow^{k_1} rR + \dots \quad \alpha = 2 \\ \\ \searrow_{k_2} sS + \dots \quad \beta = 1 \end{array}$$

$$-\frac{dC_A}{dt} = k_1 C_A^2 + k_2 C_A = (k_1 C_A + k_2) C_A \tag{5.29}$$

$$\frac{dC_R}{dt} = k_1 C_A^2 \tag{5.30}$$

$$\frac{dC_S}{dt} = k_2 C_A \tag{5.31}$$

The value of k_2 can be determined with Eq. (5.29), separating variables and integrating by partial fractions:

$$-\int_{C_{Ao}}^{C_A} \frac{dC_A}{(k_1 C_A + k_2) C_A} = \int_0^t dt$$

$$-\ln\left(\frac{k_1 C_A + k_2}{k_1 C_{Ao} + k_2}\right) + \ln\left(\frac{C_{Ao}}{C_A}\right) = k_2 t \tag{5.32}$$

The representation of $ln(C_{Ao}/C_A)$ versus t gives a straight line, whose slope is the value of k_2 as illustrated in Figure 5.6. The value of the other reaction rate coefficient cannot be calculated from the value of the intercept.

The calculation of k_1 can be done with two approaches: (1) integral method (iterative approach) and (2) differential method (direct approach).

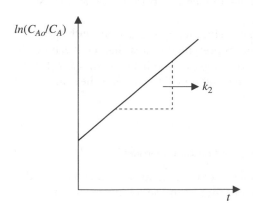

$ln(C_{Ao}/C_A)$

k_2

t

Figure 5.6 Calculation of k_2 for simultaneous reactions with combined orders.

5.2.2.1 Integral Method

It is required to evaluate the yield of R or S. For instance, for yield of S, Eq. (5.31) is divided by Eq. (5.29), and separating variables and integrating:

$$\frac{dC_S/dt}{-dC_A/dt} = -\frac{k_2 C_A}{C_A(k_1 C_A + k_2)} = -\frac{k_2}{k_1 C_A + k_2}$$

$$\int_{C_{So}}^{C_S} dC_S = -k_2 \int_{C_{Ao}}^{C_A} \frac{dC_A}{k_1 C_A + k_2}$$

$$C_S - C_{So} = -\frac{k_2}{k_1}\ln\left(\frac{k_1 C_A + k_2}{k_1 C_{Ao} + k_2}\right) = \frac{k_1}{k_2}\ln\left(\frac{k_1 C_{Ao} + k_2}{k_1 C_A + k_2}\right)$$

$$= \frac{1}{K}\ln\left(\frac{C_{Ao} + 1/K}{C_A + 1/K}\right)$$

Dividing between C_{Ao} to obtain the yield of S:

$$Y_S = \frac{C_S}{C_{Ao}} = \frac{C_{So}}{C_{Ao}} + \frac{1}{KC_{Ao}}\ln\left(\frac{C_{Ao} + 1/K}{C_A + 1/K}\right) \tag{5.33}$$

Similarly, for the yield of R:

$$Y_R = \frac{C_R}{C_{Ao}} = \frac{C_{Ro}}{C_{Ao}} + \left(1 - \frac{C_A}{C_{Ao}}\right) + \frac{1}{KC_{Ao}}\ln\left(\frac{C_A + 1/K}{C_{Ao} + 1/K}\right) \tag{5.34}$$

where

$$K = \frac{k_1}{k_2} \tag{5.35}$$

The iterative approach to evaluate k_1 is the following:

1) Assume an initial value of K.
2) Calculate the yields of S ($Y_S{}^{calc}$) and R ($Y_R{}^{calc}$) for each concentration of A using Eqs. (5.33) and (5.34), respectively
3) Calculate the following objective function (OF) based on the sum of residual values, defined as the absolute difference between calculated and experimental yields:

$$OF = \sum_{i=1}^{N}\left\{abs\left[(Y_R^{calc})_i - (Y_R^{exp})_i\right] + abs\left[(Y_S^{calc})_i - (Y_S^{exp})_i\right]\right\}$$

where N is the number of experimental data.

4) If the value of *OF* is lower than a convergence criterion (ξ), the assumed value of *K* is correct; otherwise, return to step 1 and assume a new value of *K*.
5) Since k_2 has been already calculated with Eq. (5.32), the value of k_1 is determined with Eq. (5.35).

5.2.2.2 Differential Method

Eqs. (5.29) and (5.30) are divided to obtain an expression free of time, as follows:

$$\frac{dC_R / dt}{-dC_A / dt} = -\frac{dC_R}{dC_A} = \frac{k_1 C_A^2 + k_2 C_A}{k_1 C_A^2}$$

$$-\frac{dC_R}{dC_A} = 1 + \left(\frac{k_2}{k_1}\right)\frac{1}{C_A} \tag{5.36}$$

Eq. (5.36) can be solved by means of the different approaches of differentiation described in Chapter 2. For instance, Eq. (5.36) can be transformed into:

$$-\frac{\Delta C_R}{\Delta C_A} = 1 + \left(\frac{k_2}{k_1}\right)\frac{1}{C_{Ap}} \tag{5.37}$$

where

$$\Delta C_R = |C_{R_{i-1}} - C_{R_i}|$$

$$\Delta C_A = |C_{A_{i-1}} - C_{A_i}|$$

$$C_{Ap} = \frac{C_{Ai} + C_{Ai+1}}{2}$$

The ratio of k_2/k_1 is evaluated with Eq. (5.37) as illustrated in Figure 5.7. With this ratio, the value of k_1 is calculated with Eq. (5.35).

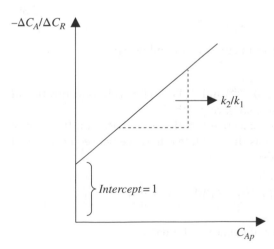

Figure 5.7 Calculation of k_2/k_1 for consecutive reactions with combined orders with the differential method.

It is important to mention that the application of the differential method requires sufficient experimental data to obtain a good fit.

5.3 Consecutive or In-Series Irreversible Reactions

5.3.1 Consecutive Reactions with the Same Order

The simplest reaction that occurs with consecutive mechanisms is the transformation of A into R, then R into S and so on, all the reactions following the same order n:

$$aA \xrightarrow{k_1} rR \xrightarrow{k_2} sS \xrightarrow{k_3} \ldots \xrightarrow{k_{n-1}} yY \xrightarrow{k_n} zZ$$

The reaction rate equations for each component are:

$$(-r_A) = -\frac{dC_A}{dt} = k_1 C_A^n \tag{5.38}$$

$$(r_R) = \frac{dC_R}{dt} = k_1 C_A^n - k_2 C_R^n \tag{5.39}$$

$$(r_S) = \frac{dC_S}{dt} = k_2 C_R^n - k_3 C_S^n \tag{5.40}$$

$$(r_Z) = \frac{dC_Z}{dt} = k_n C_Y^n \tag{5.41}$$

As an example, Figure 5.8 shows the profiles of concentrations for the case of a reaction that involves the production of R and S ($aA \xrightarrow{k_1} rR \xrightarrow{k_2} sS$).

The value of k_1 can be obtained in a similar way as the case of irreversible reactions with one component with Eq. (5.38):

For $n = 1$:

$$\ln\frac{C_{Ao}}{C_A} = k_1 t \tag{5.42}$$

For $n \neq 1$:

$$C_A^{1-n} - C_{Ao}^{1-n} = (n-1)k_1 t \tag{5.43}$$

The concentration and yield of A are evaluated as follows, for instance for $n = 1$:

$$C_A = C_{Ao}e^{-k_1 t} \tag{5.44}$$

$$Y_A = \frac{C_A}{C_{Ao}} = e^{-k_1 t} \tag{5.45}$$

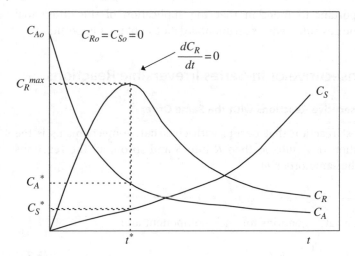

Figure 5.8 Profiles of concentration for consecutive irreversible reactions.

To evaluate k_2, the following two approaches are used.

Method 1   Using Eq. (5.39) in the point of maximum concentration of R. Eq. (5.39) is equal to zero in the point of C_R^{max}:

$$(r_R) = \frac{dC_R}{dt} = k_1 \left(C_A^* \right)^n - k_2 \left(C_R^{max} \right)^n = 0$$

where C_A^* is the concentration of A in the point of C_R^{max} (Figure 5.8). Therefore, k_1/k_2 is calculated with:

$$K = \frac{k_1}{k_2} = \left(\frac{C_R^{max}}{C_A^*} \right)^n \tag{5.46}$$

Since the values of n and k_1 have been already calculated with Eqs. (5.42) and (5.43), the value of k_2 is obtained with Eq. (5.46).

$$k_2 = k_1 \left(\frac{C_A^*}{C_R^{max}} \right)^n$$

It is important to mention that this method requires the exact value of the maximum concentration of R.

Method 2   Use of Eq. (5.39) in integrated form.
This method requires the integration of Eq. (5.39), in which the concentration of A is included. This concentration can be obtained from Eqs. (5.42) or (5.43) depending on the reaction order. Then, it is substituted in Eq. (5.39).

For instance, for $n = 1$, Eq. (5.44) is substituted in Eq. (5.39):

$$\frac{dC_R}{dt} = k_1 C_A - k_2 C_R = k_1 C_{Ao} e^{-k_1 t} - k_2 C_R$$

which can be written as a linear differential equation of first order:

$$\frac{dC_R}{dt} + k_2 C_R = k_1 C_{Ao} e^{-k_1 t}$$

which is equal to:

$$\frac{dy}{dx} + Py = Q$$

where

$$y = C_R$$
$$x = t$$
$$P = k_2$$
$$Q = k_1 C_{Ao} e^{-k_1 t}$$

The solution of this equation is the following:

$$y e^{\int P dx} = \int Q e^{\int P dx} dx + C$$

which in terms of concentrations is:

$$C_R e^{\int k_2 dt} = \int k_1 C_{Ao} e^{-k_1 t} e^{\int k_2 dt} dt + C$$

Rearranging and integrating:

$$C_R e^{k_2 t} = k_1 C_{Ao} \int e^{(k_2 - k_1)t} dt + C$$

$$C_R e^{k_2 t} = \left(\frac{k_1 C_{Ao}}{k_2 - k_1} \right) e^{(k_2 - k_1)t} + C$$

The evaluation of the integration constant C requires an initial condition. For instance, at $t = 0$, $C_R = C_{Ro}$:

$$C = C_{Ro} - \frac{k_1 C_{Ao}}{k_2 - k_1}$$

Substituting the value of C in the integrated equation:

$$C_R e^{k_2 t} = \frac{k_1 C_{Ao}}{k_2 - k_1} \left[e^{(k_2 - k_1)t} - 1 \right] + C_{Ro}$$

Finally, the concentration and yield of R are:

$$C_R = \frac{k_1 C_{Ao}}{k_2 - k_1} \left[e^{-k_1 t} - e^{-k_2 t} \right] + C_{Ro} e^{-k_2 t} \tag{5.47}$$

$$Y_R = \frac{C_R}{C_{Ao}} = \frac{k_1}{k_2 - k_1} \left[e^{-k_1 t} - e^{-k_2 t} \right] + \frac{C_{Ro}}{C_{Ao}} e^{-k_2 t} \tag{5.48}$$

In Eq. (5.48), if the initial concentrations C_{Ao} and C_{Ro} are known, and also the variation of concentration of R with respect to time, and k_1 that was evaluated with Eq. (5.42), the only unknown is k_2.

The following iterative approach to evaluate k_2 is followed:

1) Assume an initial value of k_2.
2) Calculate the yield of R (Y_R^{calc}) for each time using Eq. (5.48).
3) Calculate the following objective function (*OF*) based on the sum of residual values, defined as the absolute difference between calculated and experimental yields:

$$OF = \sum_{i=1}^{N} abs \left[\left(Y_R^{calc} \right)_i - \left(Y_R^{exp} \right)_i \right]$$

where N is the number of experimental data.
4) If the value of *OF* is lower than a convergence criterion (ξ), the assumed value of k_2 is correct; otherwise, return to step 1 and assume a new value of k_2.

For reaction orders different from one, the integration of Eq. (5.39) is more complicated, and Method 1 is more appropriate.

When the final product of the reaction is S ($aA \xrightarrow{k_1} rR \xrightarrow{k_2} sS$), its concentration is obtained by mass balance:

$$C_S = C_{Ao} - C_A - C_R \quad \text{For} \quad C_{Ro} = C_{So} = 0 \tag{5.49}$$

Substituting Eqs. (5.44) and (5.47) in Eq. (5.49):

$$C_S = C_{Ao} - C_A - C_R = C_{Ao} - C_{Ao} e^{-k_1 t} - \frac{k_1 C_{Ao}}{k_2 - k_1} \left[e^{-k_1 t} - e^{-k_2 t} \right]$$

$$C_S = C_{Ao} \left(1 - e^{-k_1 t} - \frac{k_1 C_{Ao}}{k_2 - k_1} e^{-k_1 t} + \frac{k_1 C_{Ao}}{k_2 - k_1} e^{-k_2 t} \right)$$

$$= C_{Ao} \left[1 - e^{-k_1 t} \left(1 + \frac{k_1}{k_2 - k_1} \right) + \frac{k_1 C_{Ao}}{k_2 - k_1} e^{-k_2 t} \right]$$

$$C_S = C_{Ao} \left(1 - \frac{k_2}{k_2 - k_1} e^{-k_1 t} + \frac{k_1 C_{Ao}}{k_2 - k_1} e^{-k_2 t} \right) \tag{5.50}$$

5.3.1.1 Calculation of C_R^{max} and t^*

As can be seen in Figure 5.8, the product R shows a maximum value (C_R^{max}) at the time t^*. In this point, the variation of such a concentration with respect to time is zero, so that $dC_R/dt = 0$. Then, deriving Eq. (5.47) and equalling to zero give:

$$\frac{dC_R}{dt} = \frac{k_1 C_{Ao}}{k_2 - k_1}\left[-k_1 e^{-k_1 t^*} + k_2 e^{-k_2 t^*}\right] - k_2 C_{Ro} e^{-k_2 t^*} = 0$$

When R is not present in the feed (i.e. $C_{Ro} = 0$), the time at maximum concentration of R is obtained:

$$-k_1 e^{-k_1 t^*} + k_2 e^{-k_2 t^*} = 0$$
$$k_1 e^{-k_1 t^*} = k_2 e^{-k_2 t^*}$$
$$\frac{k_2}{k_1} = e^{(k_2 - k_1)t^*}$$
$$t^* = \frac{\ln^{k_2}/k_1}{k_2 - k_1} \quad \text{For } C_{Ro} = 0 \tag{5.51}$$

Substituting t^* in Eq. (5.47) for $C_{Ro} = 0$:

$$C_R^{max} = \frac{k_1 C_{Ao}}{k_2 - k_1}\left[e^{-\frac{k_1}{k_2-k_1}\ln^{k_2}/k_1} - e^{-\frac{k_2}{k_2-k_1}\ln^{k_2}/k_1}\right]$$
$$= \frac{k_1 C_{Ao}}{k_2 - k_1}\left[\left(\frac{k_2}{k_1}\right)^{-\frac{k_1}{k_2-k_1}} - \left(\frac{k_2}{k_1}\right)^{-\frac{k_2}{k_2-k_1}}\right]$$

Factoring the term $\left(\frac{k_2}{k_1}\right)^{-\frac{k_1}{k_2-k_1}}$:

$$C_R^{max} = \left(\frac{k_1 C_{Ao}}{k_2 - k_1}\right)\left(\frac{k_2}{k_1}\right)^{-\frac{k_1}{k_2-k_1}}\left[1 - \left(\frac{k_2}{k_1}\right)^{-\frac{k_1-k_2}{k_2-k_1}}\right]$$
$$= \left(\frac{k_1 C_{Ao}}{k_2 - k_1}\right)\left(\frac{k_2}{k_1}\right)^{-\frac{k_1}{k_2-k_1}}\left[1 - \frac{k_1}{k_2}\right] = \left(\frac{k_1 C_{Ao}}{k_2 - k_1}\right)\left(\frac{k_2}{k_1}\right)^{-\frac{k_1}{k_2-k_1}}\left[\frac{k_2-k_1}{k_2}\right]$$
$$C_R^{max} = \left(\frac{k_1 C_{Ao}}{k_2}\right)\left(\frac{k_2}{k_1}\right)^{-\frac{k_1}{k_2-k_1}} = \left(\frac{k_1 C_{Ao}}{k_2}\right)\left(\frac{k_1}{k_2}\right)^{\frac{k_1}{k_2-k_1}} = C_{Ao}\left(\frac{k_1}{k_2}\right)^{1+\frac{k_1}{k_2-k_1}}$$

And, finally:

$$C_R^{max} = C_{Ao}\left(\frac{k_1}{k_2}\right)^{\frac{k_2}{k_2-k_1}} \tag{5.52}$$

This equation also can be easily derived by substitution of t^* (Eq. 5.51) in Eq. (5.44) for $n = 1$, assuming that $t = t^*$ and $C_A = C_A^*$:

$$C_A^* = C_{Ao}e^{-k_1 t^*} = C_{Ao}e^{-k_1 \frac{\ln^{k_2}/k_1}{k_2 - k_1}} = C_{Ao}e^{\frac{k_1}{k_2 - k_1}\ln^{k_2}/k_1}$$

$$= C_{Ao}e^{\ln\left(\frac{k_2}{k_1}\right) - \frac{k_1}{k_2 - k_1}} = C_{Ao}\left(\frac{k_2}{k_1}\right)^{-\frac{k_1}{k_2 - k_1}}$$

C_R^{max} is obtained from Eq. (5.46) for $n = 1$, and substituting C_A^*:

$$C_R^{max} = C_A^*\left(\frac{k_1}{k_2}\right) = C_{Ao}\left(\frac{k_2}{k_1}\right)^{-\frac{k_1}{k_2 - k_1}}\left(\frac{k_1}{k_2}\right)$$

$$= C_{Ao}\left(\frac{k_1}{k_2}\right)^{\frac{k_1}{k_2 - k_1}}\left(\frac{k_1}{k_2}\right) = C_{Ao}\left(\frac{k_1}{k_2}\right)^{\frac{k_1}{k_2 - k_1} + 1} = C_{Ao}\left(\frac{k_1}{k_2}\right)^{\frac{k_2}{k_2 - k_1}}$$

5.3.1.2 Calculation of C_R^{max} and t^* for $k_1 = k_2$

In the particular case of $k_1 = k_2$, Eqs. (5.51) and (5.52) are indeterminate. Therefore, it is necessary to apply L'Hôpital's rule, which postulates that:

$$\lim_{x \to x_0}\frac{f(x)}{g(x)} = \lim_{x \to x_0}\frac{f'(x)}{g'(x)} \tag{5.53}$$

To use this rule, logarithms are applied to Eq. (5.52):

$$\ln\frac{C_R^{max}}{C_{Ao}} = \ln\left(\frac{k_1}{k_2}\right)^{\frac{k_2}{k_2 - k_1}} = \frac{k_2}{k_2 - k_1}\ln\left(\frac{k_1}{k_2}\right) = \frac{\ln\left(\frac{k_1}{k_2}\right)}{(k_2 - k_1)\Big/k_2} \tag{5.54}$$

In this equation:

$$f(x) = \ln\left(\frac{k_1}{k_2}\right)$$

$$g(x) = (k_2 - k_1)\Big/k_2$$

The corresponding derivatives with respect to k_1 are:

$$f'(x) = \frac{1}{k_1}$$

$$g'(x) = -\frac{1}{k_2}$$

Applying Eq. (5.53) in (5.54) using the previous values:

$$\lim_{k_1 \to k_2} \frac{\ln\left(\frac{k_1}{k_2}\right)}{(k_2-k_1)\big/k_2} = \lim_{k_1 \to k_2} \frac{1\big/k_1}{-1\big/k_2} = \lim_{k_1 \to k_2} -\frac{k_2}{k_1} = -1$$

Therefore, Eq. (5.54) is:

$$\ln \frac{C_R^{max}}{C_{Ao}} = -1$$

And, finally, the maximum concentration C_R^{max} when $k_1 = k_2$ is:

$$\frac{C_R^{max}}{C_{Ao}} = 0.3679 \qquad (5.55)$$

Similarly, when applying L'Hôpital's rule in Eq. (5.51), the following expression is obtained to calculate t^* when $k_1 = k_2$:

$$t^* = \frac{1}{k_1} = \frac{1}{k_2} \qquad (5.56)$$

Example 5.3 The following reaction in series was studied at 12 °C, and the experimental results reported in Table 5.3 were obtained. Find the kinetic model.

$$A \xrightarrow{k_1} R \xrightarrow{k_2} S$$

Solution

Assuming that both reactions are carried out with first-order kinetics, from Eq. (5.42) it is possible to evaluate k_1. For the first data:

$$k_1 = \frac{1}{t}\ln\frac{C_{Ao}}{C_A} = \frac{1}{4}\ln\left(\frac{0.8000}{0.5948}\right) = 7.409 \times 10^{-2}$$

Table 5.3 Data and results of Example 5.3.

Time (min)	C_A (mol/lt)	C_R (mol/lt)	$k_1 \times 10^2$	Y_R^{exp}	Y_R^{calc} ($k_2 = 0.01$)	Y_R^{calc} ($k_2 = 0.02$)	Y_R^{calc} ($k_2 = 0.0135$)
0	0.8000	0.000	–	0.000	0.000	0.000	0.000
4	0.5948	0.1995	7.409	0.249	0.249	0.245	0.248
12	0.3288	0.4299	7.409	0.537	0548	0.513	0.535
28	0.1056	0.5475	7.232	0.684	0.727	0.610	0.0683
60	0.0094	0.4237	7.407	0.529	0.621	0.397	0.0530
Average	–	–	7.364×10^{-2}	OF	0.1474	0.2353	0.0043

The complete results are present in Table 5.3. It is seen that the values of k_1 are constant, with an average value of 7.364×10^{-2}. The evaluation of k_2 can be done with the two methods described previously in this chapter:

Method 1   Using Eq. (5.39) in the point of maximum concentration of R. From Table 5.3, it is observed that the maximum concentration of R is $C_R^{max} = 0.5475$ mol/lt, and at this point the concentration of A is $C_A^* = 0.1056$ mol/lt. Applying Eq. (5.46) for $n = 1$:

$$K = \frac{k_1}{k_2} = \left(\frac{C_R^{max}}{C_A^*}\right)^n = \frac{0.5475}{0.1056} = 5.185$$

$$k_2 = \frac{k_1}{K} = \frac{7.364 \times 10^{-2}}{5.185} = 1.420 \times 10^{-2}$$

Method 2   Using Eq. (5.39) in integrated form.
This method requires the yields of R ($Y_R = C_R/C_{A0}$), which are calculated and summarized in Table 5.3.

According to this method, it is first necessary to assume an initial value of k_2 and calculate the yields of R (Y_R^{calc}) with Eq. (5.48) for each time with the already known value of k_1. Then, the absolute error is calculated with the experimental and calculated yields of R:

$$Absolute\ error = \sum_{i=1}^{N} abs\left[\left(Y_R^{calc}\right)_i - \left(Y_R^{exp}\right)_i\right]$$

When the absolute error is minimum, the assumed value of k_2 is correct.

Table 5.3 presents the calculation for different values of k_2. Additionally, for the absolute errors for different values of k_2, the minimum value is found at 0.0135. This value differs slightly from that found with Method 1 (0.0142); however, Method 2 gives a better prediction since the calculation involves minimization of the difference between the experimental and calculated yields.

5.3.2   Consecutive Reactions with Combined Orders

In these cases, the disappearance of A follows a kinetics of order n_1, and the disappearance of R follows a kinetics of order n_2, where $n_1 \neq n_2$. If $n_1 = n_2$, the equations developed in the previous section will apply.

$$aA \xrightarrow{\substack{n_1 \\ k_1}} rR \xrightarrow{\substack{n_2 \\ k_2}} sS$$

The reaction rate equations for each component are:

$$(-r_A) = -\frac{dC_A}{dt} = k_1 C_A^{n_1} \tag{5.57}$$

$$(r_R) = \frac{dC_R}{dt} = k_1 C_A^{n_1} - k_2 C_R^{n_2} \tag{5.58}$$

$$(r_S) = \frac{dC_S}{dt} = k_2 C_R^{n_2} \tag{5.59}$$

The value of k_1 can be calculated from Eq. (5.57):

For $n_1 = 1$:

$$\ln \frac{C_{Ao}}{C_A} = k_1 t \tag{5.60}$$

For $n_1 \neq 1$:

$$C_A^{1-n_1} - C_{Ao}^{1-n_1} = (n_1 - 1) k_1 t \tag{5.61}$$

The yield of A is obtained directly from Eqs. (5.60) or (5.61). For instance, for a reaction order of 2:

$$Y_A = \frac{C_A}{C_{Ao}} = \frac{1}{1 + k_1 C_{Ao} t} \quad \text{For } n_1 = 2 \tag{5.62}$$

The evaluation of k_2 can be done by the following procedure:

1) Assume a value of n_2.
2) Calculate k_2 from Eq. (5.58) in the point of maximum concentration R (C_R^{max}). The values of k_1 and n_1 are calculated from Eq. (5.60) or (5.61).

$$k_2 = k_1 \frac{\left(C_A^* \right)^{n_1}}{\left(C_R^{max} \right)^{n_2}} \tag{5.63}$$

where $C_A{}^*$ is the concentration in the point of maximum concentration of A (C_R^{max}).
3) Knowing k_1, k_2, n_1, n_2 and the concentration of A (C_A) at different times, the concentration of R (C_R) can be calculated by means of the numerical solution of the differential equation given by Eq. (5.58).
4) Evaluate the objective function (OF) based on the sum of the differences between experimental and calculated concentrations of R.

$$OF = \sum_{i=1}^{N} abs \left[\left(C_R^{calc} \right)_i - \left(C_R^{exp} \right)_i \right]$$

where N is the number of experimental data.

5) If the value of OF is lower than a convergence criterion (ξ), the assumed value of n_2 is correct; otherwise, return to step 1 and assume a new value of n_2.

Example 5.4 In the following reaction, the product R is obtained by dimerization and further isomerization. The experimental data reported in Table 5.4 were obtained when the following reaction was conducted in liquid phase. Find the rate equation for this reaction.

$$2A \xrightarrow{k_1} R \xrightarrow{k_2} S$$

Solution

Assume that the first reaction is elemental, that is, the disappearance of A follows a kinetics of order two ($n_1 = 2$). Then, k_1 is obtained directly from Eq. (5.61). For the first data:

$$k_1 = \frac{1}{t}\left(\frac{1}{C_A} - \frac{1}{C_{Ao}}\right) = \frac{1}{0.03}\left(\frac{1}{0.76} - \frac{1}{1}\right) = 10.526$$

The complete results are summarized in Table 5.4. It is observed that the values of k_1 are constant, so that the first reaction follows a second-order kinetics with an average value of $k_1 = 10.0835$. To determine the order of the second reaction (n_2) and k_2, it is first necessary to know the values of C_R^{max} and C_A^*, which from Table 5.4 are:

Table 5.4 Data and results of Example 5.4.

Time (min)	C_A (mol/lt)	C_R (mol/lt)	k_1	$C_R^{calc.}$	$\lvert C_R^{calc.} - C_R^{exp.} \rvert$
0	1.00	0.000	–	0.000	–
0.03	0.76	0.173	10.526	0.172	0.001
0.06	0.63	0.207	9.788	0.204	0.003
0.10	0.51	0.184	9.608	0.179	0.005
0.15	0.39	0.134	10.427	0.130	0.004
0.20	0.33	0.094	10.152	0.091	0.003
0.30	0.25	0.048	10.000	0.047	0.001
Average	–	–	10.0835	OF	0.017

$$C_R{}^{max} = 0.207\,\text{mol/lt}$$

$$C_A{}^* = 0.63\,\text{mol/lt}$$

Assuming that $n_2 = 1$, the value of k_2 is determined with Eq. (5.61):

$$k_2 = k_1 \frac{\left(C_A^*\right)^{n_1}}{\left(C_R^{max}\right)^{n_2}} = k_1 \frac{\left(C_A^*\right)^2}{C_R^{max}} = (10.0835)\frac{(0.63)^2}{0.207} = 19.334$$

So that Eq. (5.58) is:

$$\frac{dC_R}{dt} = 10.0835 C_A^2 - 19.334 C_R \tag{5.64}$$

where C_A can be calculated with Eq. (5.62):

$$C_A = \frac{C_{Ao}}{1 + k_1 C_{Ao}t} = \frac{1}{1 + 10.0835 t}$$

Finally, the differential equation to be solved is:

$$\frac{dC_R}{dt} = \frac{10.0835}{\left(1 + 10.0835 t\right)^2} - 19.334 C_R$$

The solution of this equation can be done with the fourth-order Runge–Kutta method, and the results are presented in Table 5.4. It is observed that the calculated values of C_R are similar to the experimental ones, with an average error of 0.017. This indicates that the proposed kinetic model [Eq. (5.64)] is in good agreement with the experimental values.

6

Special Topics in Kinetic Modelling

This chapter is devoted to the discussion of the following special topics in kinetic modelling:

- *Data reconciliation*: Experiments conducted in a bench-scale heavy oil catalytic hydrotreating unit were used to illustrate the application of a data reconciliation approach in order to minimize the inconsistencies of mass balances due to experimental errors.
- *Sensitivity analysis of parameters*: The role of sensitivity analysis during kinetic parameter estimation is reported. An approach consisting of various steps – initialization of parameter values, non-linear parameter estimation and parameter sensitivity analysis – is applied to assure that kinetic parameters are properly estimated and the convergence of the objective function to the global minimum is achieved. The method is illustrated with experimental data reported in the literature for the hydrodesulphurization of benzothiophene.
- *Kinetics of enzymatic reactions*: Experimental data reported in the literature for different enzymatic reactions were used to determine the kinetic parameters of the Michaelis–Menten equation. Parameter estimation was performed by employing linear regression and graphic, integral and non-linear regression methods.
- *Kinetics of catalytic cracking reaction*: A simple method to estimate gasoline, gas and coke yields in the catalytic cracking process is discussed. The method only requires experimental information about the variation with time of products yields, which are correlated using a fifth-order polynomial.
- *Hydrodesulphurization of petroleum distillates*: It was carried out in a batch reactor using a commercial catalyst and reaction conditions similar to those of industrial practice. Reaction orders and activation energies for each feed were determined by two approaches (linear and non-linear regression).

Chemical Reaction Kinetics: Concepts, Methods and Case Studies, First Edition.
Jorge Ancheta.
© 2017 John Wiley & Sons Ltd. Published 2017 by John Wiley & Sons Ltd.

Although some of the cases of study in this chapter are for heterogeneous reactions, the kinetic modelling can be done with all the methods and equations described in Chapters 1 through 5. In other words, the treatment of the kinetic data can be performed as pseudo-homogeneous models. Even though there is a catalyst present in some of the reactions, based on some simplifications, the approaches can be used for estimating the kinetic parameters by relating the reaction rate expressions in such a way that the terms involving the catalyst are eliminated.

6.1 Data Reconciliation

The data generated in different reactor scales (laboratory, bench-scale, pilot plant, semi-industrial or commercial) are used in various stages of process development to perform mass and energy balances, kinetic studies, reactor modelling and process simulation, among others. The accuracy of the collected data strongly affects such calculations in such a way that if mass balance closure is not properly achieved, a series of assumptions are needed (e.g. values normalization) to compensate for the lack of preciseness.

The quality of the data depends on several factors, among them: (1) the scale of the experimentation: small-scale setups frequently have problems with sample handling; (2) repeatability of analysis: variations in reaction conditions affect the composition of products sent to gas chromatography; (3) flowrate variations: in some cases, commercial units exhibit large changes in the feed flowrate with the consequent impact on product production; and (4) instruments calibration: if proper care is not taken, the flowrates would be wrongly registered.

For a situation in which there is uncertainty of the data, not all is lost: there are suitable approaches that can help take advantage of the available information.

These methods are required to cope with common data-processing problems such as measurement uncertainties, lack of measurement availability for critical variables, limited knowledge of the process behaviour and information redundancy in the available measurements.

The data obtained in an experiment usually contain mistakes, which can be random or bulk. These mistakes can seriously affect the results and cause false conclusions when the information is analysed. That is the reason why data reconciliation and outlier determination are used.

In most of the research reports in the literature, these aspects are not properly taken into consideration, and the authors either use the experimental data as such or normalize them without distinguishing the source of the error. This practice, although common, is not the best approach to generate high-quality experimental information.

In the literature, there are some reports on the application of data reconciliation. Examples include experimental kinetics (Phillips and Harrison, 1993), refinery heat exchanger networks (Smaïli *et al.*, 2001), crude oil distillation units (Basak *et al.*, 2002), gas pipeline systems (Bagajewicz and Cabrera, 2003), distillation control (Wen Li *et al.*, 2006), coking plants (Hu and Shao, 2006), the natural gas industry (Oliveira and Aguiar, 2009), catalytic distillation (Buchaly *et al.*, 2012) and fluid catalytic cracking (Pinheiro *et al.*, 2012). These are certainly the few reports that one can find about data reconciliation examples; however, none of them give enough details on the application of the method, or a step-by step description that allows one to properly use this approach.

This section aims at presenting a detailed description of the use of data reconciliation techniques. Data collected from a bench-scale unit for hydrotreating of heavy oils are used as examples.

6.1.1 Data Reconciliation Method

Process data may contain inaccurate information, because the measurements are obtained with instruments and signals that intrinsically have measurement error sources.

The error term is composed of random error and gross error. The *random error* is originated by factors that affect the measurement of a variable in a random way. This type of error adds variability to data, but it does not affect the average development of the group.

The *gross error* occurs by deviations of the instruments that are consistently erroneous (bad calibration), fouling of the measurement dispositive and non-random events that affect the process, such as process leaks. This error tends to be positive or negative consistently.

Generally, the measurements with gross errors will lead to severe incorrect process information. Gross error detection is an important aspect in process data validation.

The basis of data reconciliation is a model that makes use simultaneously of process models and measurements. Data reconciliation is the estimation of the process variables using the information contained in the process measurements and models. Data reconciliation adjusts the measurements to be consistent with the balances.

In general, the optimal estimates of the process variables by data reconciliation are solutions to least squares or maximum likelihood restrictions. In practice, this criterion is sufficiently powerful for improving data at industrial or even laboratory scales. The simplest problem of data reconciliation occurs in the reconciliation of process flow of a plant.

When the constraints and the measurement equations are linear, the solution of the reconciliation problem can be developed analytically.

However, in non-linear cases, it is normally impossible to derive an explicit expression for the reconciled states. Various methods are possible, depending on the approach selected to handle the constraints and on the optimization technique used to minimize the criterion. Tong and Crowe (1995) detected gross error in data reconciliation by principal component analysis. Wang and Romagnoli (2003) presented a framework for robust data reconciliation based on a generalized objective function. Wang *et al.* (2004) proposed an improved measured test–nodal test (MT-NT) method. Miao *et al.* (2009) applied a support vector regression approach to data reconciliation and gross error detection. Maronna and Arcas (2009) proposed a method based on regression analysis. Zhang *et al.* (2010) made data reconciliation with a quasi-weighted least squares estimator.

On the other hand, from a graphical point of view, an outlier is an observation that falls a long distance from other values in a random sample. A boxplot is a useful tool to describe the data behaviour, which was introduced by Tukey (1997). It uses the upper quartile (Q_3) and lower quartile (Q_1) (defined as the 25th and 75th percentiles, respectively). The difference between quartiles is called the *interquartile range (IQ)*.

A boxplot is constructed by drawing a box between the lower and upper quartiles with a solid line through the box to locate the median. The following quantities (fences) are needed to identify the extreme values in the tail of the distribution:

Lower inner fence (LIF) : $Q_1 - 1.5\,IQ$
Superior inner fence (SIF) : $Q_3 + 1.5\,IQ$
Lower outer fence (LOF) : $Q_1 - 3\,IQ$
Superior outer fence (SOF) : $Q_3 + 3\,IQ$

A point beyond an inner fence on either side is considered a mild outlier. A point beyond an outer fence is considered an extreme outlier.

6.1.2 Results and Discussion

6.1.2.1 Source of Data

A series of experiments was conducted in a bench-scale catalytic hydro-treatment of a heavy oil unit at constant reaction conditions. For the purpose of explaining the application of the data reconciliation method, it does not matter if the reaction is catalytic or not, since the catalyst will not be taken into consideration, but only the results of the mass balances.

The heavy oil showed the following properties: 13° API (American Petroleum Institute) gravity, 5.25 wt% sulphur, 83.28 wt% carbon, 0.49 wt% nitrogen and 10.91 wt% hydrogen. The gas feedstock was pure hydrogen.

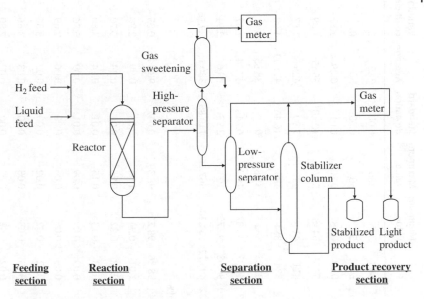

Feeding section   Reaction section   Separation section   Product recovery section

Figure 6.1 Bench-scale plant flow sheet.

The bench-scale unit consists of four sections: the feeding section, reaction section, separation section and product recovery section; a simplified flow sheet is shown in Figure 6.1. Liquid is fed to the reactor from a container with a heating system to maintain the feed in liquid state. The inlet and outlet gas flows are measured using mass flow meters, while the liquid is measured in weight by a digital scale. The liquid feed is pre-mixed with hydrogen, and both are fed to the reactor. The products of reaction are sent to a series of separators that operate at different conditions of pressure and temperature to separate gas and liquid products. Once the system is at steady state, samples of products are obtained by weight balance at regular time intervals (12 h). The gaseous product is analysed by gas chromatography. The gas feed is pure hydrogen. The liquid feed composition is obtained by elemental analysis.

To perform the data reconciliation analysis, a data series with significant variations in some process variables was selected. These data were obtained at a temperature of 385 °C, liquid hourly space velocity (*LHSV*) of 0.29 h^{-1}, pressure of 50 kg/cm^2 and hydrogen-to-oil ratio of 5000 scf/Bbl. Table 6.1 shows the experimental data. The average value, minimum value, maximum value, standard deviation value and variation coefficient (standard deviation between the average value multiplied by 100) were determined. The values in weight represent the total amount of liquid and gases at the entrance and at the exit of the reactor collected in the duration of the balance. The control of reactor temperature and pressure is accurate, and there is no variation in the values obtained.

Table 6.1 Experimental data of liquid feed and product.

Balance no.	1	2	3	4	5	6	7	8	9	Minimum value	Maximum value	Standard deviation	Average	Variation coefficient
Temperature, °C	385	385	385	385	385	385	385	385	385	385	385	0.00	385	0
Pressure, atm	50	50	50	50	50	50	50	50	50	50	50	0.00	50	0
$LHSV$, h^{-1}	0.30	0.30	0.29	0.29	0.29	0.29	0.29	0.29	0.28	0.28	0.30	0.01	0.29	1.87
H$_2$/HC ratio ft^3/bbl	5430	5430	5570	5506	5570	5623	5584	5596	5774	5430	5774	105.01	5565	1.89
Liquid feed, g	437	437	426	431	426	422	425	424	411	411	437	7.99	426.56	1.87
Liquid product, g	435	422	395	420	402	422	400	420	384	384	435	16.46	411.11	4.00
Input gas, g	38.85	38.85	38.85	38.85	38.85	38.85	38.85	38.85	38.85	38.85	38.85	0	38.85	0
Gas product, g	46.73	46.46	48.36	47.38	48.26	47.16	47.67	46.73	46.37	46.37	48.36	0.74	47.24	1.57
Liquid yield, vol%	102.97	99.78	95.75	100.75	97.63	103.45	97.24	102.49	96.68	95.75	103.45	2.93	99.64	2.94
H$_2$ consumption, ft^3/Bbl	402.11	567.16	242.57	349.21	234.48	376.82	312.06	441.32	495.72	234.48	567.16	110.85	380.16	29.16
Gas composition, mol%														
Hydrogen	98.39	98.24	98.36	98.36	98.37	98.37	98.35	98.34	98.36	98.24	98.39	0.047	98.34	0.05
H$_2$S	0.94	1.03	0.98	0.97	0.97	0.98	0.98	1.01	1.01	0.94	1.03	0.03	0.99	2.96
Methane	0.24	0.26	0.23	0.24	0.23	0.23	0.26	0.23	0.22	0.22	0.26	0.01	0.24	5.70
Ethane	0.13	0.15	0.14	0.14	0.14	0.14	0.13	0.14	0.13	0.13	0.15	0.01	0.14	3.60
Propane	0.09	0.10	0.09	0.09	0.09	0.09	0.09	0.09	0.09	0.09	0.09	0.01	0.09	5.86
n-Butane	0.06	0.06	0.06	0.06	0.06	0.06	0.06	0.06	0.06	0.06	0.06	0.00	0.06	2.69
i-Butane	0.02	0.02	0.02	0.02	0.02	0.02	0.02	0.02	0.02	0.02	0.02	0.00	0.02	3.36
n-Pentane	0.03	0.03	0.03	0.03	0.03	0.03	0.03	0.03	0.03	0.03	0.03	0.00	0.03	3.66
i-Pentane	0.02	0.02	0.02	0.02	0.02	0.02	0.02	0.02	0.02	0.02	0.02	0.00	0.02	2.80
C$_5^=$/C$_6^+$	0.08	0.09	0.07	0.07	0.07	0.06	.06	0.06	0.06	0.06	0.09	0.01	0.07	14.69

It is seen that the gas weight at the entrance of the reactor is constant (38.85 g), which is due to the high precision of the gas flowmeter used. The gas product weight exhibited a variation coefficient of 1.57%. The measurement of the outlet gas is not as accurate as that of the inlet gas because the produced hydrocracked light gases during reaction cause disturbances of flowrates. In any case, the values do not differ significantly. However, the variation of the weight of the gas product, together with the variation of the weight of the liquid feed, has a great influence on the determination of H_2 consumption, which presents the highest variation coefficient (29.16%). The hydrogen content in the gas product is almost constant (variation coefficient of 0.05%). The inlet and outlet liquid streams reported high variation coefficients. This behaviour is common in small-scale units since handling and weighting of the liquid samples become more inaccurate when dealing with heavy viscous hydrocarbons, such as those used in our experiments. This behaviour of the liquid feed has a straight effect on the LHSV calculation, which has the same variation coefficient as the liquid feed (1.87%). The molar composition of the gas, except hydrogen and $C_5^=/C_6^+$, has variation coefficients between about 2.69% and 5.86%. The $C_5^=/C_6^+$ are the hydrocarbons with the largest variation coefficient (14.69%). The differences in gas composition in all balances are somewhat normal and are attributed to small variations in reaction conditions, particularly space–velocity.

6.1.2.2 Global Mass Balances

Global and individual mass balances are first developed by means of the following equations.

Global balance:

$$S_1 + S_2 - S_3 - S_4 = 0 \tag{6.1}$$

Hydrogen balance:

$$S_1 y_{1H2} + S_2 y_{2H2} - S_3 y_{3H2} - S_4 y_{4H2} = 0 \tag{6.2}$$

Sulphur balance:

$$S_1 y_{1S} + S_2 y_{2S} - S_3 y_{3S} - S_4 y_{4S} = 0 \tag{6.3}$$

Carbon balance:

$$S_1 y_{1C} + S_2 y_{2C} - S_3 y_{3C} - S_4 y_{4C} = 0 \tag{6.4}$$

The weight fractions of carbon (y_{1C}), hydrogen (y_{1H2}) and sulphur (y_{1S}) in the liquid feed were those determined with the elemental analysis. Gas feed is composed only of hydrogen, so that $y_{2S} = y_{2C} = 0$ and $y_{2H2} = 1$. The weight fractions of carbon (y_{4C}), hydrogen (y_{4H2}) and sulphur (y_{4S}) in the gas product (S_4) were obtained with the molar composition from

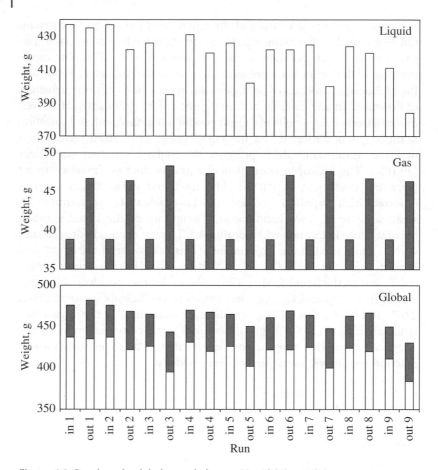

Figure 6.2 Bench-scale global mass balances. Liquid (□), gas (■).

the chromatographic analysis, the molecular weight of the gases and the atomic weight of each element.

Because there is not information about the elemental analysis of liquid products, only the global mass balance of the data can be obtained. However, even though there are unmeasured variables, the reconciliation data technique can estimate the values of these variables as reported by Vasebi *et al.* (2011). Figure 6.2 presents the mass balances. As can be seen, the global balances are not satisfied in any case; and, in some cases, there is more mass at the outlet than at the inlet of the bench-scale unit. These variations are owing to the digital balance precision. To have better quality data and properly satisfy the mass balances, data reconciliation is required.

Table 6.2 Required information to determine outliers.

	Minimum value	1st quartile	2nd quartile	3rd quartile	Maximum value	LOF	LIF	SIF	SOF
Amount, g									
Liquid feed	411	424	426	431	437	403	413.5	441.5	452
Liquid product	384	400	420	422	435	334	367	455	488
Gas feed	38.85	38.85	38.85	38.85	38.85	38.85	38.85	38.85	38.85
Exit gas	46.37	46.73	47.16	49.67	48.36	43.91	45.32	49.08	50.49
Gas composition, mol%									
Hydrogen	98.14	98.19	98.21	98.31	98.37	97.83	98.01	98.49	98.67
H_2S	0.94	0.97	0.97	1.01	1.03	0.84	0.90	1.07	1.14
Methane	0.23	0.23	0.23	0.24	0.26	0.19	0.21	0.26	0.28
Ethane	0.13	0.13	0.14	0.14	0.15	0.12	0.13	0.15	0.15
Propane	0.09	0.09	0.09	0.09	0.10	0.08	0.08	0.10	0.10
i-Butane	0.02	0.02	0.02	0.02	0.02	0.02	0.02	0.02	0.02
n-Butane	0.06	0.06	0.06	0.06	0.06	0.05	0.06	0.06	0.06
n-Pentane	0.03	0.03	0.03	0.03	0.03	0.02	0.03	0.03	0.03
i-Pentane	0.02	0.02	0.02	0.02	0.02	0.02	0.02	0.02	0.02
$C_5^=/C_6^+$	0.06	0.06	0.07	0.07	0.09	0.04	0.05	0.08	0.09

6.1.2.3 Outlier Determination

Before any calculation, definition about whether the information is between acceptable limits is needed, that is, the outlier existence must be verified. The information analysis of the case of study was performed using the box tool.

As mentioned in this chapter, the limits of the boxplot are determined by calculating the fences, and with the minimum and maximum values it can be identified if there are experimental data outside of the boxplot (i.e. an outlier). The results of the application of this method are presented in Table 6.2, and the boxplots of the cases with an outlier are shown in Figure 6.3. It can be observed that the composition of $C_5^=/C_6^+$ hydrocarbons and propane in gas product in run 2 have experimental data that are considered an outlier, and they are not taken into account for the data reconciliation.

6.1.2.4 Data Reconciliation

As mentioned in this chapter, the most reliable experimental flow measurement is the feed gas stream. The exit gas stream has also small variation (variation coefficient of 1.57%). The liquid product stream has greater

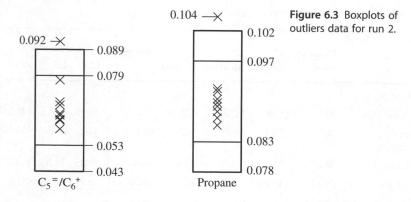

Figure 6.3 Boxplots of outliers data for run 2.

variation than the feed product stream. To simplify the problem, the data reconciliation method requires that some of the experimental results are to be considered as accurate and do not require any modification. Therefore, based on the aforementioned, the mass and composition of the feed gas stream, the composition of the liquid feed and the composition of inlet gas are assumed to be true values, while the mass of feed and product liquid streams, the composition of the liquid product stream and the gas product stream are to be corrected.

To analyse the experimental information, two cases of study were developed:

Case 1: The amounts of liquid feed, liquid product, gas product and elemental composition of the liquid product can be modified.

Case 2: Since the gas product stream has little variation, only the feed and product liquids streams and the elemental composition of the liquid product can be modified.

The objective function of the data reconciliation problem is the minimization of the square error between experimental and calculated values of the four streams entering and exiting the reactor (liquid and gases), divided by the standard deviation of the experimental data, subject to global and individual mass balances of hydrogen, carbon and sulphur. The optimization model is:

$$Min\ OF = \frac{\left(S_{1exp} - S_{1calc}\right)^2}{\sigma_{S1exp}} + \frac{\left(S_{2exp} - S_{2calc}\right)^2}{\sigma_{S2exp}} + \frac{\left(S_{3exp} - S_{3calc}\right)^2}{\sigma_{S3exp}}$$
$$+ \frac{\left(S_{4exp} - S_{4calc}\right)^2}{\sigma_{S4exp}} + \left(y_{3H2exp} - y_{3calc}\right)^2 + \left(y_{3Cexp} - y_{3calc}\right)^2$$
$$+ \left(y_{3Sexp} - y_{3Scalc}\right)^2$$

$$(6.5)$$

Subject to:

Global balance:

$$S_{1calc} + S_{2calc} - S_{3calc} - S_{4calc} = 0 \tag{6.6}$$

Hydrogen balance:

$$S_{1calc}y_{1H2} + S_{2calc}y_{2H2} - S_{3calc}y_{3H2} - S_{4calc}y_{4H2} = 0 \tag{6.7}$$

Carbon balance:

$$S_{1calc}y_{1C} + S_{2calc}y_{2C} - S_{3calc}y_{3C} - S_{4calc}y_{4C} = 0 \tag{6.8}$$

Sulphur balance:

$$S_{1calc}y_{1S} + S_{2calc}y_{2S} - S_{3calc}y_{3S} - S_{4calc}y_{4S} = 0 \tag{6.9}$$

The optimization model is a non-lineal system with equality lineal restrictions.

6.1.2.5 Analysis of Results

The optimization model was implemented and solved in MS Excel, using the Solver tool.

6.1.2.5.1 Case 1

Figure 6.4 shows the results for Case 1. After the reconciliation, the global and individual mass balances are met in all cases. Before the data reconciliation can be done, the amount of mass of product liquid has a difference between the maximum and minimum value of 51 g (Table 6.1). After the reconciliation data, this difference is lower (~30 g). In the liquid feed streams, the variation of the same differences is not so great (30.8 g after the reconciliation, and 26 g before the reconciliation). In the gas product, there is slight variation in these differences before and after the reconciliation (~2 g). This behaviour indicates that this variable is not meaningful to data reconciliation and can be maintained as fixed, as is done in Case 2.

About the mass balance by element (C, H and S), there is no way to compare them before and after the reconciliation, because, as was mentioned in this chapter, the experimental liquid product composition was not obtained.

The amount of carbon in the liquid product is always lower than that of the feed, as can be seen in Figure 6.4. The amount of carbon in the gas product is similar in all the runs (between 3.4 and 4.2 g).

Similar observations can be done for H_2 and S. The liquid feed stream contains less H_2 than the liquid product stream, which is the result of the hydrocarbon hydrogenation. Finally, the decrease of the amount of

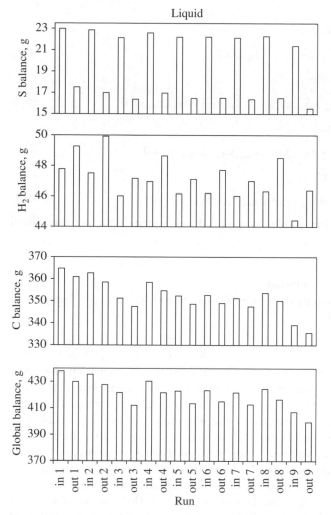

Figure 6.4 Reconciled mass balances, Case 1. Liquid (□), gas (■).

sulphur in the liquid product stream indicates the degree of desulphuri-
zation of the liquid feed stream. Having more or less the same amount of
each element in the product stream is a result of the operation at constant
conditions.

6.1.2.5.2 *Case 2*
As in Case 1, the global and individual mass balances are met after the
reconciliation procedure. The results of Case 2 are presented in Figure 6.5.

The effect of not considering the mass of gas product as an optimization
variable has little effect on the results of the reconciliation data. Negligible

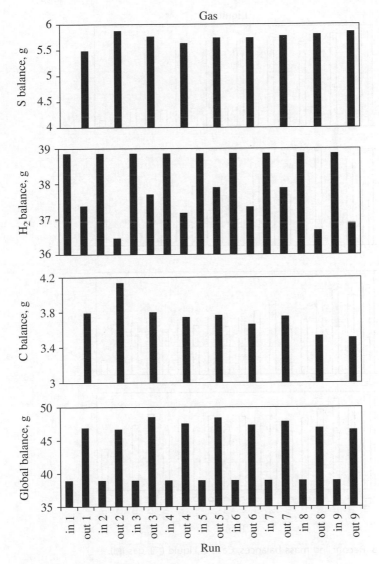

Figure 6.4 (*Continued*)

variations in hydrogen, sulphur and carbon balances as well as in the global balances with respect to Case 1 were observed.

Although the variation coefficient of the gas product stream is similar to that of the liquid feed stream, the values in mass are different. The mass of the liquid feed stream is about ten times that of the gas product stream, so that, when varying the gas product stream, the effect on the mass balance is insignificant.

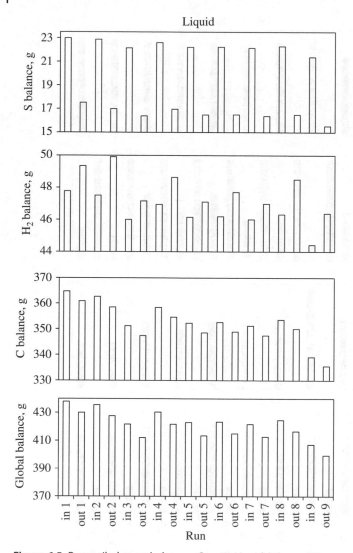

Figure 6.5 Reconciled mass balances, Case 2. Liquid (☐), gas (■).

6.1.2.5.3 *Comparison with Traditional Normalization*

The use of average values and normalizations of data either by distributing the differences between inlet and outlet streams or by using rules of thumb based on experience are the most common approaches to close the mass balances. Table 6.3 compares the results of these three methods compared with those obtained with the reconciliation data for run 3. It is important to mention that, as there is no information about the

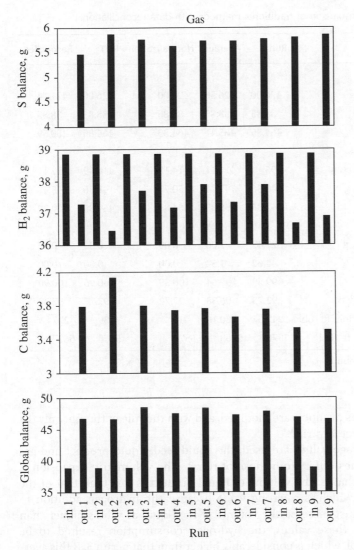

Figure 6.5 (*Continued*)

composition of the liquid product, the individual mass balances are not calculated for the traditional normalization methods.

The average values calculated from the experimental data do not meet the mass balance (1.52% of loss), while for all the methods the mass balance is close to 100%. For the methods of normalization and rule of thumb, the amount of liquid feed is considered correct (426 g) and the amount of liquid product is variable. In the normalization method, both

Table 6.3 Comparison of traditional methods with data reconciliation.

	Run 3	Average	Normalization	ROT	Reconciled
Inlet					
Liquid feed, g	426.00	426.56	426.00	426.00	421.74
Gas feed, g	38.85	38.85	38.85	38.85	38.85
Total, g	464.85	465.41	464.85	464.85	460.59
Outlet					
Liquid product, g	395	411.11	414.15	416.49	412.11
Gas product, g	48.36	47.24	50.70	48.36	48.48
Total, g	443.36	458.35	464.85	464.85	460.59
Balance					
Loss of mass, g	21.49	7.06	0.00	0.00	0.00
Loss of mass, %	4.62	1.52	0.00	0.00	0.00
Liquid yield, vol%	99.59	99.64	100.39	100.96	100.90
Liquid yield, wt%	92.73	96.38	97.22	97.77	97.72
H_2 consumption, ft^3/Bbl	242.57	380.16	−40	239	227
H_2/HC ratio, ft^3/Bbl	5570	5563	5570	5570	5627

ROT, Rule of thumb.

liquid and gas products are modified, and with the rule of thumb, only the liquid product is varied.

If the average values are used, the liquid feed, liquid product and gas product values differ from those of run 3. The volumetric expansion is 99.64 vol%, that is, the volume of liquid product is less than that of the liquid feed, which is a wrong value because the hydrogen addition to the liquid must increase the volume of the liquid at values higher than 100%. With these values, the hydrogen consumption resulted to be 380.16 ft^3/Bbl, which is considerably higher than that of run 3. If this average value is used to calculate the hydrogen requirement for a process design, the hydrogen plant capacity would be overestimated.

To meet the mass balance, in the normalization method the amount of product streams are modified based on the amount of liquid feed and total mass at the entrance. This implies that both product streams, liquid and gas, are proportionally increased. The volumetric expansion achieves an adequate value, being greater than 100%, but the increase in the mass of the gas product stream without considering the individual mass balances increases the amount of hydrogen in the gas product stream, making the hydrogen consumption negative (i.e. it seems that the process produces hydrogen instead of consuming it). This is a wrong result in a hydrotreating experiment since it would not be possible to determine the hydrogen requirement and thus to design the process.

In the case of rule of thumb, from previous experience it is known that the liquid product is the stream that needs to be modified (increased) to meet the mass balance. This assumption is made on the fact that working with heavy viscous samples causes problems when handling and sampling the liquid oil, so that the losses are most probably due to this. With this modification, the liquid product stream volume is greater than that of the liquid feed stream (volumetric expansion of 100.96%).

With the data reconciliation, the global mass balance and simultaneously the individual mass balances are met, modifying the liquid product composition and the mass balance of the streams. The variation of the mass quantities is maintained at minimum because the modifications are done by minimizing the difference between the experimental data and the optimum values.

The hydrogen consumption and the H_2–oil ratio are very similar to those of the experimental data of run 3. In addition, volumetric product yield is higher than 100%, as expected.

The advantages of the reconciliation data method are that it permits an estimate of unmeasured values and that the variation of the values considered as variables respecting the original values is minimized.

6.1.3 Conclusions

A reconciliation technique was performed on data generated in a bench-scale catalytic hydrotreatment of a heavy oil unit. The method indicates that it is highly important to combine global and individual mass balances to correct the experimental data, because the modification of the process stream values to meet the global mass balance changes the individual mass balances in a significant manner.

The boxplot outlier procedure is a suitable approach to analyse the experimental data and decide which values can be eliminated to improve the quality of experimental information. The appropriate definition of the variables that are considered correct is crucial for the optimization model to obtain accurate results with the reconciliation approach. The reconciliation data permit estimating unmeasured variables and minimizing the variation of the modified values.

The normalization and the rule of thumb approaches only assure the closure of the global mass balance if unmeasured values exist in the data.

6.2   Methodology for Sensitivity Analysis of Parameters

Most of the mathematical models used for representing any type of phenomena (or situation) occurring not only during chemical reactions but also in other areas (e.g. thermodynamics, environmental sciences, molecular modelling etc.) involve parameters that need to be estimated from experimental data. The models can be supported on theoretical, semi-theoretical/semi-empirical or empirical bases, and their parameters can have theoretical meaning or be simply correlation constants. Linear regression is the most widely used method for parameter estimation due to its simplicity and easy manner to interpret the results (i.e. by representing the data in 2-dimensional (2D) plots and examining how the experimental points deviate from the straight line). The common way that researchers often express the strength of the relationship between two variables is by the correlation coefficient (r) or determination coefficient (correlation coefficient squared, r^2), concepts from statistics that are used to see how well trends in the predicted values follow trends in experimental values, and range between 0 and 1. If there is no relationship between the predicted values and the experimental ones $r = 0$, $r^2 = 0$ or is very low. As the strength of the relationship between the predicted values and experimental values increases, so does the correlation coefficient. A perfect fit gives a coefficient of 1. Thus, the higher the correlation and determination coefficients, the better the fit (Chapra and Canale, 1990).

Sometimes, a model can be transformed into different linear equations, such as in the case of the Michaellis–Menten (M-M) model used in enzymatic kinetics, as can be seen in Table 6.4 (Quastel and Woolf, 1926; Lineweaver and Burk, 1934; Eadie, 1942; Christensen and Palmer, 1980). It has been reported that the values of parameters of the M-M model calculated with the linear equations given in Table 6.4 can be slightly different, and it is recommended and more accurate to use that model which when representing the data in a 2D plot gives better distribution of the experimental points along the straight line. This has been confirmed by non-linear regression analysis (Avery, 1983).

Another example of difficulties when using linear regression analysis to estimate kinetic parameters has been reported recently. For studying kinetics of hydrocracking of heavy oils in a perfectly mixed continuous reaction system, some authors transform the resulting reaction rate equations into various straight lines, as is presented in Table 6.5, and then they calculate separately the values of each parameter, k_0, k_1 and k_2 (Callejas and Martínez, 1999). By this way, the condition $k_0 = k_1 + k_2$ is not satisfied,

Table 6.4 Example of different linear equations obtained from a same model.

Method	Equation
Michaellis–Menten Model	$v = \dfrac{V_{max}[S]}{K_m + [S]}$
Lineweaver and Burk	$\dfrac{1}{v} = \dfrac{1}{V_{max}} + \dfrac{K_m}{V_{max}}\dfrac{1}{[S]}$
Eadie–Hofstee	$\dfrac{v}{[S]} = \dfrac{V_{max}}{K_m} - \left(\dfrac{1}{K_m}\right)v$
Augustinsson	$v = V_{max} - K_m \dfrac{v}{[S]}$
Woolf	$\dfrac{[S]}{v} = \dfrac{K_m}{V_{max}} + \dfrac{1}{V_{max}}[S]$
Nonlinear regression	$SSE = \sum |(v^c - v^e)|^2 = \sum\limits_{i=1}^{N} \left| \dfrac{V_{max}\, C_s^e}{K_m + C_s^e} - v^e \right|$

Table 6.5 Example of parameter estimation with linear and non-linear regression analyses.

Kinetic model	
	$Feed(A) \xrightarrow{k_1} Light\ Oils(B)$
	$Feed(A) \xrightarrow{k_2} Gases(C)$
	$(r_A) = -(k_1 + k_2)C_A = -k_0 C_A$
	$(r_B) = k_1 C_A$
	$(r_C) = k_2 C_A$
Linear regression analysis (k_0, k_1 and k_2 are determined independently)	$\left(\dfrac{C_{Ao} - C_A}{C_A}\right) = k_0 \left(\dfrac{1}{WHSV}\right)$
	$\left(\dfrac{C_{Ao} - C_A}{C_A}\right)\left(\dfrac{C_B - C_{Bo}}{C_A}\right) = k_1 \left(\dfrac{1}{WHSV}\right)$
	$\left[\dfrac{(C_{Ao} - C_A)C_C}{C_{Ao} C_A}\right] = k_2 \left(\dfrac{1}{WHSV}\right)$
Nonlinear regression analysis (k_0, k_1 and k_2 are determined simultaneously)	$SSE = \sum\limits_{i=1}^{3}\left(C_i^{calc} - C_i^{exp}\right)^2$
	C_i^{calc} evaluated with the above equations

while when determining these kinetic parameters simultaneously by non-linear regression this situation is not presented. The error between experimental and calculated yields has been shown to be lower with parameter values determined with the latter approach (Ancheyta *et al.*, 2005).

As can be observed from the examples described here, linear regression analysis can sometimes present problems when estimating parameters for a given model. That is why non-linear regression is a more common approach when modelling heterogeneous kinetic systems, in which the main objective is to optimize the values of the model parameters that provide the best fit to the experimental data. This non-linear parameter estimation is carried out by using the least squares method, searching the best set of parameters that minimizes the sum of squared errors (SSE) between measured and calculated values.

When using non-linear regression for parameter estimation, the task turns into a non-linear optimization problem, which can be solved by optimization methods (Marquardt, 1963), such as Gauss–Newton and Levenberg–Marquardt, among others. The Levenberg–Marquardt method is, of course, the most popular alternative to the Gauss–Newton method of finding the minimum of a function that is a sum of squares of non-linear functions. Some models, such as those used for describing heterogeneous kinetics, can have several parameters (sometimes hundreds) to be estimated, and be highly non-linear; in those cases, when determining the values of parameters, multiple solutions of the objective function during the optimization process (i.e. multiple minima) can be obtained, and the best set of parameters is not guaranteed. The optimal solution depends mostly on the initial guess of parameters (Seferlis and Hrymak, 1996; Varma *et al.*, 1999).

Most of the kinetics studies reported in the literature only give parameter values, r, r^2, residuals (differences between experimental and calculated values), absolute errors and/or *SSE*s without enough evidence to assure that parameter values correspond to the global minima of the objective function, and consequently the model accuracy is not clearly established. The sensitivity analysis is a tool that allows for validating the values of parameters obtained by regression analysis. Sensitivity analysis is a way to assure that the solution of the objective function with a given set of parameters does correspond to the global minimum and not to local minima in the parameter optimization process.

The objective of this section is to describe an approach based on sensitivity analysis to determine the best set of parameter values during the parameters optimization process. The procedure is exemplified with parameter estimation of a kinetic model and experimental data of hydrodesulphurization of benzothiophene reported in the literature (Kilanowski and Gates, 1980).

6.2.1 Description of the Method

A direct and universal approach that can be a guarantee of the best solution during a parameter estimation process is not easy to develop.

The main difficulties when estimating parameters in heterogeneous kinetics are:

1) The complexity of the model, which can be from simple algebraic equations to complex differential equation systems, linear or highly non-linear in nature
2) The source and precision of experimental data, which can come from the literature (from one or more references) or from the researcher's own or literature experiments specially designed to perform kinetic studies in which all the care has been put to assure a kinetic regime
3) The robustness of the optimization algorithm; in most of the cases, the Levenberg–Marquardt method is used since it has been shown to be superior over others (Reklaitis *et al.*, 1983).
4) The numerical method employed for solving the model equations; for instance, orthogonal collocation has been reported to fail for dynamic simulation of plug flow packed bed reactors, and the method of characteristics has been preferred.

Therefore, what is presented here is not such a magic method but an approach that takes into consideration various steps (e.g. initialization of parameter values, non-linear parameter estimation and parameter sensitivity analysis) to determine and validate the set of parameters that minimizes the differences between experimental and calculated experimental values. A schematic representation of the proposed methodology is shown in Figure 6.6.

6.2.1.1 Initialization of Parameters
The optimal solution during non-linear parameter estimation depends mostly on the initial guess of parameters values. The initialization of parameters is a problem frequently found in non-linear estimation that may converge to local minima and not to the global minimum during the parameter optimization process.

If the kinetic model and the corresponding parameters have been reported previously by other authors, no matter the differences in reaction conditions, catalyst, feed, reaction system and so on, at least the order of magnitude of the reported parameters values should be employed as an initial guess. If there are not reported values, an iterative analysis of orders of magnitude of the parameters should be performed. This approach can sometimes be very tedious, since it implies the evaluation of the objective function for different sets of parameters, starting, say, with $k_i = 1$ (where k is the parameter to be estimated, and $i = 1$ is N parameters). Then, the value of each parameter is changed one at a time – say, $k_j = 10$ – keeping constant the values of the others ($k_i = 1$, for $i \neq j$), and the objective function is evaluated again. For any modification in the value of k_i, the change

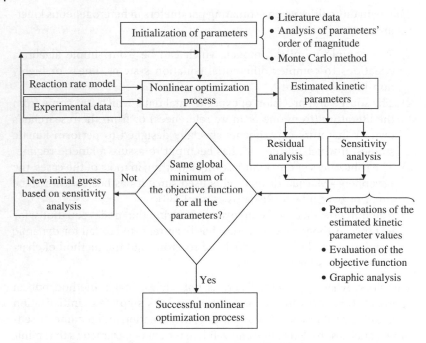

Figure 6.6 Proposed methodology for parameter estimation.

in the objective function is examined, and the influence of each parameter on the objective function (sensitivity of each parameter) is determined.

This procedure is repeated for different values of $k_i < 1$ (0.1, 0.01, 0.001 etc.) and/or $k_i > 1$ (10, 100, 1000 etc.) as many times as necessary. A lower value of the objective function indicates the correct order of magnitude of the parameter value. This means that if one parameter is changed (e.g. from 1 to 10), and the value of the objective function increases, it is more likely that this parameter value is <1.

This approach allows for determining the order of magnitude of the parameter values and makes easier the estimation of initial values. Of course, it requires patience and certain expertise in its use.

An initial guess of parameter values can also be obtained using the Monte Carlo method (Rubinstein, 1981), which consists mainly of the following steps: (1) an initial guess of parameters is determined using random numbers; (2) with this initial guess of parameters, the objective function (e.g. the *SSE*) is calculated; and (3) this procedure is repeated M times ($M > 1000$), and the minimum of the M values of the sum of squares of residuals is determined. The set of initial guess of parameters that corresponds to this minimum can be used as initial estimates in the non-linear parameter optimization process.

6.2.1.2 Non-linear Parameter Estimation

The reliable solution of non-linear parameter estimation is an important computational problem when modelling heterogeneous kinetic systems. This non-linear parameter estimation is commonly carried out by using the least squares method in order to find the global minimum of the following objective function:

$$SSE = \sum_{i=1}^{N\,data} \left(y_{exp} - y_{calc}\right)^2 \qquad (6.10)$$

The Marquardt (or Levenberg–Marquardt) method uses the method of linear descent in early iterations and then gradually switches to the Gauss–Newton approach. Most of the scientific software (the so-called "solvers") uses the Marquardt method for performing non-linear regression analysis.

Most often, non-linear regression is done without weighting, giving equal weight to all points (as in Eq. 6.11), as is appropriate when experimental scatter is expected to be the same in all parts of the curve. If experimental scatter is expected to vary along the curve, then the points can be weighted differentially. The most frequently used weighting method is called "weighting by $1/y^2$" and is expressed as follows:

$$SSE = \sum_{i=1}^{N\,data} \frac{1}{y_{exp}^2} \left(y_{exp} - y_{calc}\right)^2 \qquad (6.11)$$

Sometimes, the data come with additional information about which points are more reliable. For example, different data may correspond to averages of different numbers of experimental trials; in this case, weighting of the data should be added in the objective function to obtain better estimates:

$$SSE = \sum_{i=1}^{N\,data} w_i \left(y_{exp} - y_{calc}\right)^2 \qquad (6.12)$$

where w_i is a weighting factor.

6.2.1.3 Sensitivity Analysis

Sensitivity analysis is commonly employed to assess that in the non-linear parameter estimation, the set of parameters does correspond to the global minimum and not to local minima. Sensitivity analysis is applied to each of the estimated parameters by means of perturbations of the parameter value (keeping the other parameters in their estimated values). Perturbations are preferably done in the range of ±20%. For each perturbation in the parameter values, the objective function is re-evaluated, and then for each parameter the perturbation percentage is plotted against the corresponding value of the objective function. If all perturbations in all the parameters give the minimum of the objective function with their original values (0% perturbation), then the global minimum has been achieved. On the contrary, if at least one parameter does not give the same minimum as the others at 0% perturbation, that means poor non-linear parameter estimation. To correct this, the values of the wrong estimated parameters are re-determined by examining the sensitivity plot, then parameter sensitivity is carried out again on these parameters, and now, finally, the global minimum is guaranteed.

6.2.1.4 Residual Analysis

Analysis of residual distribution, calculated as the difference between experimental and predicted values, is frequently practiced by some authors as a way to demonstrate that the estimated parameters for a given model accurately predict the experimental values. Plots of the residuals are used to check the quality of the fit. Graphical analysis of the residuals is the single most important technique for determining the need for model parameters refinement.[19] A plot of residual values against the number of experimental observations is commonly used, and a regular distribution of residuals with no prediction bias should be observed, thus proving the adequacy of the proposed model and the calculated parameters. On the contrary, if there is a pattern, the parameters are wrongly estimated.

Residual analysis is certainly useful to see graphically the precision of estimations. However, it cannot guarantee by itself the achievement of the global minimum of the objective function. It is better to use both residual analysis and sensitivity analysis to assure that parameters are properly estimated.

6.2.2 Results and Discussion

6.2.2.1 Experimental Data and the Reaction Rate Model from the Literature

Experimental data reported by Kilanowski and Gates (1980) were employed to illustrate the application of the proposed methodology for parameter estimation. These data correspond to the hydrodesulphurization of benzothiophene conducted in a steady-state differential-flow microreactor containing particles of sulphided CoMo–Al_2O_3 catalyst. The study was carried out at reaction temperatures of 252.5 °C, 302 °C, and 332.5 °C, and partial pressures in the following ranges: benzothiophene (BT), 0.015–0.23 atm; H_2, 0.20–2.0 atm; and H_2S, 0.02–0.14 atm. Differential conversion data were obtained to determine reaction rates directly. Catalyst deactivation was negligible over hundreds of hours of operation. A summary of experimental results is shown in Table 6.6.

Table 6.6 Summary of results of benzothiophene HDS kinetics (taken from Kilanowski and Gates, 1980).

T (°C)	P_{BT} (atm)	P_{H2} (atm)	P_{H2S} (atm)	P_{He} (atm)	$r_{BT} \times 10^7$ (gmol/g$_{cat}$ s)
252.5	0.105	0.215	0.022	0.819	0.272
252.5	0.068	0.234	0.026	1.04	0.309
252.5	0.197	0.842	0.020	–	1.04
252.5	0.063	1.14	0.124	–	1.10
252.5	0.063	1.14	0.094	–	1.19
252.5	0.114	0.985	0.021	–	1.25
252.5	0.063	1.11	0.045	–	1.29
252.5	0.082	1.07	0.022	–	1.37
252.5	0.064	1.14	0.023	–	1.42
252.5	0.053	1.2	0.025	–	1.46
252.5	0.046	1.26	0.025	–	1.51
252.5	0.015	1.86	0.038	–	1.77
252.5	0.018	1.94	0.040	–	2.29
252.5	0.022	2.04	0.041	–	2.55
252.5	0.024	1.82	0.037	–	2.62
252.5	0.036	1.39	0.028	–	2.70
252.5	0.030	1.54	0.032	–	3.03
252.5	0.026	1.67	0.034	–	3.15
302	0.118	0.238	0.025	0.920	1.84
302	0.080	0.272	0.030	1.24	2.00
302	0.071	1.28	0.142	–	4.71
302	0.070	1.26	0.101	–	5.48
302	0.015	1.86	0.038	–	5.51
302	0.026	1.67	0.034	–	6.54
302	0.073	1.33	0.054	–	7.41
302	0.231	0.836	0.023	–	7.85
302	0.212	0.866	0.022	–	8.1
302	0.155	0.971	0.024	–	8.22
302	0.125	1.05	0.024	–	8.43
302	0.043	1.79	0.036	–	8.78
302	0.106	1.11	0.025	–	8.84
302	0.074	1.35	0.027	–	9.16

(Continued)

Table 6.6 (Continued)

T (°C)	P_{BT} (atm)	P_{H2} (atm)	P_{H2S} (atm)	P_{He} (atm)	$r_{BT} \times 10^7$ (gmol/g$_{cat}$ s)
302	0.091	1.17	0.025	–	9.33
332.5	0.127	0.268	0.024	0.980	5.35
332.5	0.014	1.86	0.038	–	12.0
332.5	0.070	1.28	0.143	–	12.2
332.5	0.069	1.26	0.102	–	14.8
332.5	0.025	1.67	0.035	–	17.0
332.5	0.072	1.33	0.055	–	20.1
332.5	0.229	0.834	0.027	–	21.6
332.5	0.210	0.863	0.026	–	22.0
332.5	0.153	0.97	0.027	–	23.0
332.5	0.043	1.79	0.037	–	23.6
332.5	0.123	1.05	0.027	–	24.6
332.5	0.073	1.35	0.029	–	25.6
332.5	0.090	1.17	0.027	–	26.1
332.5	0.104	1.11	0.027	–	26.5

Kilanowski and Gates (1980) also reported that the following Hougen–Watson kinetic models well represented the experimental information regarding hydrodesulphurization of benzothiophene:

at 252.5 °C:

$$r = \frac{kp_{BT}p_{H_2}}{\left(1 + K_{BT}p_{BT} + K_{H_2S}p_{H_2S}\right)^2} \tag{6.13}$$

at 302 °C and 332.5 °C:

$$r = \frac{kp_{BT}p_{H_2}}{\left(1 + K_{BT}p_{BT} + K_{H_2S}p_{H_2S}\right)\left(1 + K_{H_2}p_{H_2}\right)} \tag{6.14}$$

6.2.2.2 Initialization of Parameters

Although in this case the authors have estimated and reported kinetic parameter values, they were not taken into account as a first option to give a better explanation of this step, and the Monte Carlo method was employed instead.

First, an initial range of k_i values needs to be defined. A common range to be used for an initial guess is 1–100. Figure 6.7a shows the variation of

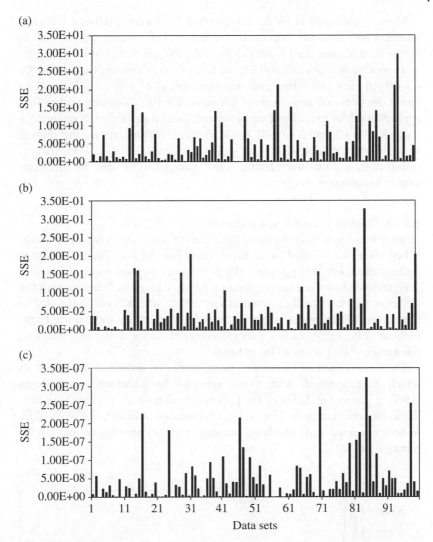

Figure 6.7 Results of Monte Carlo simulation at 252.5 °C. (a) $1 < k < 100$, (b) $0.01 < k < 10$, (c) $0.001 < k < 0.00001$.

the objective function (*SSE*) when random k_i values are employed. For illustration purposes, only 100 iterations with parameter k of the model given by Eq. (6.13) at 252.5 °C will be presented, but a similar approach was followed for K_{H2S} and K_{BT}. The lowest values of the *SSE* were found at also low k values, indicating that k is more likely to be in the direction of unity ($k \to 1$).

Then, a new range of k values is specified, for instance $0.01 < k < 10$, and another 100 random iterations are done. The results are shown in Figure 6.7b. Again, the lowest values of the *SSE* are found at low k values. This process is repeated, and Figure 6.7c presents the results for $0.00001 < k < 0.001$. For this latter case, very low values of *SSE* were found, and hence the order of magnitude of k is about 1×10^{-5}. The same procedure was followed for the other two parameters, and their orders of magnitude were: $K_{H2S} = 10$; and $K_{BT} = 10$. For the other temperatures corresponding to a different model with four kinetic parameters (Eq. 6.14), different orders of magnitude were found. These values can be used for initialization of parameters.

6.2.2.3 Results of Non-linear Estimation

Once the order of magnitude of the different parameters has been established, they can be used as an initial guess for the non-linear parameter optimization. For this purpose, the Marquardt method was employed.

Figure 6.8 shows the corresponding iterative process. It is seen that the objective function really started at low values, which is obviously due to the correct initial guess of parameters determined by the Monte Carlo method. The optimization process required about 160 iterations, and parameter values seem to be optimal.

The final results of calculated parameters are shown in Table 6.7, in which a comparison with those reported by Kilanowski and Gates (1980) is presented. Most of the parameter values are equal or quite similar to the reported ones. The highest differences, although not really high, in both reported and calculated parameter values are observed at a temperature of 302 °C.

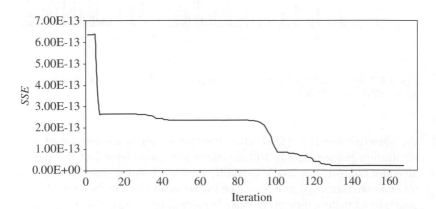

Figure 6.8 Iterative process for minimization of the objective function at 252.5 °C.

Table 6.7 Comparison of reported and calculated kinetic parameters.

Temperature (°C)	Parameter	Reported values[14]	Calculated values	Optimized values
252.5	k''	3.40×10^{-5}	3.3346×10^{-5}	3.3346×10^{-5}
	K_{BT}	3.95×10^{1}	3.9512×10^{1}	3.9512×10^{1}
	K_{H2S}	1.20×10^{1}	1.1914×10^{1}	1.13184×10^{1}
302.0	k''	2.36×10^{-4}	2.0197×10^{-4}	
	K_{H2}	1.79×10^{-1}	1.5684×10^{-1}	
	K_{BT}	2.03×10^{2}	1.7640×10^{2}	
	K_{H2S}	1.57×10^{2}	1.3699×10^{2}	
332.5	k''	2.83×10^{-4}	2.8860×10^{-4}	
	K_{H2}	6.54×10^{-2}	6.3078×10^{-2}	
	K_{BT}	8.71×10^{1}	8.8709×10^{1}	
	K_{H2S}	8.73×10^{1}	8.9922×10^{1}	

k'' (gmol/g_{cat} s atm^2), K_{BT} (atm^{-1}), K_{H2S} (atm^{-1}), K_{H2} (atm^{-1}).

6.2.2.4 Sensitivity Analysis

Analysis of parameter sensitivity was practiced for k, K_{H2S} and K_{BT} in the model at 252.5 °C (Eq. 6.13), and for k, K_{H2S}, K_{BT} and K_{H2} in the model at 302 °C and 332.5 °C (Eq. 6.14), by means of ±20% perturbations in the original parameter values determined by non-linear regression and reported earlier in this chapter. For each perturbation in each parameter, the objective function was evaluated, and the results are presented in Figures 6.9, 6.10 and 6.11 for the models at 252.5 °C, 302 °C and 332.5 °C, respectively.

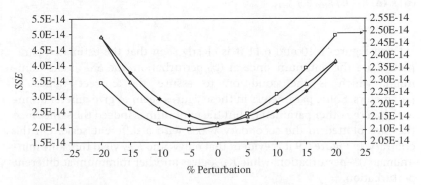

Figure 6.9 Sensitivity analysis of calculated parameters for the model at 252.5 °C. (◆) k, (△) K_{BT}, (□) K_{H2S}.

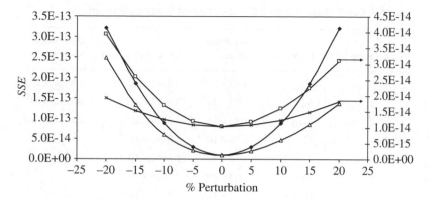

Figure 6.10 Sensitivity analysis of calculated parameters for the model at 302 °C. (◆) *k*, (△) K_{BT}, (×) K_{H2}, (□) K_{H2S}.

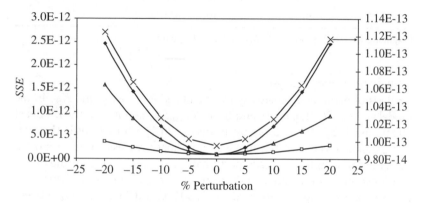

Figure 6.11 Sensitivity analysis of calculated parameters for the model at 332.5 °C. (◆) *k*, (△) K_{BT}, (×) K_{H2}, (□) K_{H2S}.

From Figures 6.10 and 6.11 it is clearly seen that the estimated parameters are the optimum since at 0% perturbation, the *SSE* is the minimum, which is the condition to assure the correct values of parameters. Some parameters in these figures seem to give different minima than the other parameters, but the minimum is indeed the same since they are plotted in the secondary *y*-axis with a different scale. On the contrary, in Figure 6.9 it is evident that only *k* and K_{BT} yield the same minimum at 0% perturbation, while K_{H2S} gives another minimum at different perturbation.

From a graphic analysis of Figure 6.9, the new value of K_{H2S} can be obtained. To find it by graphic visualization, one can try a kind of "zoom" in the perturbation percentage – say, ±10% instead of ±20% – just to expand the scale. The results with this new perturbation range are presented in Figure 6.12. As can be seen, the minimum of the objective function is achieved at –5% perturbation of the original value of K_{H2S}. The optimized value of this parameter is shown in the last column of Table 6.7.

With the optimized values of kinetic parameters, sensitivity analysis is performed again and the results are shown in Figure 6.13. Now, the three parameters gave the same minimum at 0% perturbation, and hence the global minimum is guaranteed and the optimization process is successfully finished.

It is then clear that by sensitivity analysis, one can find those parameters providing different minima than others at different perturbation values, as was demonstrated in Figures 6.9 and 6.12. In such cases, a graphic

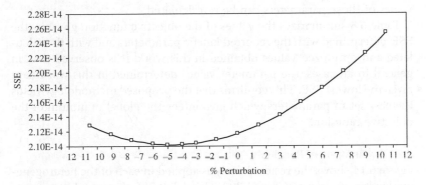

Figure 6.12 Sensitivity analysis of parameter K_{H2S} with ±10% perturbation for the model at 252.5 °C.

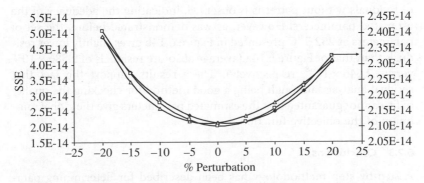

Figure 6.13 New sensitivity analysis of optimized parameters for the model at 252.5 °C. (◆) k, (△) K_{BT}, (×) K_{H2}, (□) K_{H2S}.

Table 6.8 SSE determined with reported, calculated and optimized kinetic parameters.

Temperature (°C)	Set of kinetic parameters		
	Reported values[a]	Calculated values	Optimized values
252.5	2.100×10^{-14}	2.117×10^{-14}	2.100×10^{-14}
302	1.039×10^{-14}	1.034×10^{-14}	1.034×10^{-14}
332.5	9.990×10^{-14}	9.947×10^{-14}	9.947×10^{-14}

a) Values re-calculated.

examination of the sensitivity analysis curve must be done, and with the corresponding perturbation values giving the different minima, new values of those parameters can be re-calculated.

Table 6.8 summarizes the values of the objective function given by the SSE, determined with the reported kinetic parameters and with the calculated and optimized values obtained in this work. It is observed that, in general in all cases, the parameter values determined in this work have given the lowest SSE. This confirms that the proposed methodology yields the best set of parameters which guarantees the global minimum of the objective function.

6.2.2.5 Analysis of Residuals

Figure 6.14 shows the residual analysis applied in each of the heterogeneous kinetic models at 252.5 °C, 302 °C and 332.5 °C. Figure 6.14a presents the results for the first set of calculated parameters, and Figure 6.14b the results for the optimized values of parameters, in which only the corresponding values at 252.5 °C are shown. In both cases, regular distribution of residuals without patterns is observed, indicating the adequacy of the estimated parameters. However, as was demonstrated before, the set of parameters at 252.5 °C presented in Figure 6.14b gives slightly lower residuals than that of Figure 6.14a (average absolute residuals of 2.586×10^{-8} versus 2.646×10^{-8}, respectively). These results support the fact that residual analysis, although being a good method for checking the quality of fit, cannot guarantee that the estimated parameters give the global minimum of the objective function.

6.2.3 Conclusions

A step-by-step methodology has been described for determining parameters in kinetic models. The main advantage of the method is the achievement of the optimal values of kinetic parameters by assuring

(a)

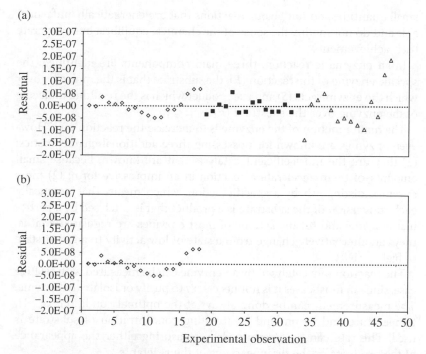

(b)

Figure 6.14 Residual analysis for the model at (◇) 252.5 °C, (■) 302 °C, (Δ) 332.5 °C.

the minimization of the objective function to the global minimum and not to local minima.

Various approaches are considered in the methodology: initialization of parameters (analysis of orders of magnitude, and Monte Carlo simulations), non-linear parameter estimation, parameter sensitivity analysis and residual analysis, which if applied properly can guarantee the best set of parameters for a given model.

Parameter estimation with the case study used showed that the methodology indeed assures the optimization of kinetic parameters values, giving lower errors than reported ones.

6.3 Methods for Determining Rate Coefficients in Enzymatic Catalysed Reactions

Enzymes are molecules produced by cells of live organisms with the specific function of catalysing chemical reactions which, without them, would proceed at a low rate. As catalysts, enzymes are strong and effective, act in small quantities, do not favour reactions that are energetically unfavour-able and do not modify the sense of the chemical equilibria but accelerate their achievement.

In an enzymatic reaction, three main components are identified: the specific enzyme of the reaction (E), the substrate that is the molecule over which the enzyme acts (S) and the product which is the resulting molecule of the enzyme over the substrate (P).

The main function of the enzyme is to increase the reaction rate. How-ever, enzymes are known for possessing three additional characteristics: (1) they are the most efficient catalysts that are known, because small amounts of them accelerate a reaction in an impressive form; (2) most enzymes distinguish by a specific action, which means that practically each conversion of the substrate is a product that is catalysed by a partic-ular enzyme; and (3) the actions of most enzymes are regulated, that is, they can alternatively change from a state of low activity to another state of high activity.

The reaction rate catalysed by an enzyme can be measured with relative ease, since in most cases it is not necessary to purify or isolate the enzyme. The measurement can be done always at the optimal conditions of pH, temperature and so on, and a saturating concentration of substrate is used. The rate can be determined by measuring either the appearance of the products or the disappearance of the reactants.

The reaction rates catalysed by enzymes are in general proportional to the first power of the enzyme concentration (they are of first order with respect to the enzyme). The rate varies linearly with the concentration of substrate at low concentrations (first order with respect to the substrate), and it becomes independent of its concentration (zero order) at high concentrations.

The model of Michaelis–Menten (Michaelis and Menten, 1913) con-tinues to be the base of the enzymatic kinetics. There are different forms to determine the rate coefficients of enzymatic reactions using this model. When experimental data on the reaction rate and concentration of the substrate are available, the rate coefficients can be calculated by means of a graphical method. Another approach to obtain these parameters is by means of linear regression, by applying the methods of Lineweaver–Burk (Lineweaver and Burk, 1934), Eadie–Hofstee (Eadie, 1942), Augus-tinsson (Avery, 1983) or Woolf (Avery, 1983), which are different linear transformations of the Michaelis–Menten model. Likewise, the rate coefficient also can be evaluated by non-linear regression using the Marquardt–Levenberg method.

For the case in which experimental data and concentration of substrate are available, the integral method is used, which is based on the integra-tion of the Michaelis–Menten equation; and the resulting equation is transformed into a linear form. In addition, a functionality between the concentration and time can be derived, and with it the values of the reaction rate can be determined, and the previous methods can be used.

Most of the literature reports calculations of the rate coefficient of the Michaelis–Menten model with the methods of Lineweaver–Burk and Eadie–Hofstee, which indicate that the latter yields better results. This section, then, aims at using experimental information from the literature to estimate the rate coefficients of the Michaelis–Menten model. A comparison is done between the four methods with the graphical, integral and non-regression approaches.

6.3.1 The Michaelis–Menten Model

6.3.1.1 Origin

The first studies of enzymatic reaction were reported by V. Henri (1902) and later by Michaelis and Menten (1913), who developed this theory and proposed a reaction rate equation that explained the kinetic behaviour of enzymes.

Michaelis and Menten (1913) formulated a simple model to explain this behaviour. It was proposed that the enzyme E can be combined in irreversible mode with the substrate S to form an intermediate complex composed of the enzyme and the substrate ES, which is further decomposed to form the products P and the free enzyme in its original form (Figure 6.15).

6.3.1.2 Development of the Model

Starting from the following reaction:

$$E + S \underset{k_{-1}}{\overset{k_1}{\rightleftharpoons}} ES \xrightarrow{k_2} P + E$$

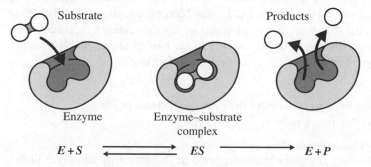

Figure 6.15 Model for the behaviour of an enzymatic reaction.

The reaction rate of the complex *ES* is:

$$\frac{d[ES]}{dt} = k_1[E][S] - (k_2 + k_{-1})[ES] \tag{6.15}$$

For a given time, the total concentration of the enzyme, E_t, is:

$$[E_t] = [E] + [ES] \tag{6.16}$$

Substituting Eq. (6.16) in Eq. (6.15) and applying the approximation of steady-state $d[ES]/dt = 0$ give:

$$r = \frac{V_{max}[S]}{K_m + [S]} \tag{6.17}$$

where:

$$K_m = \frac{k_2 + k_{-1}}{k_1} \tag{6.18}$$

$$V_{max} = k_2[E_t] \tag{6.19}$$

where K_m is the constant of Michaelis–Menten, as a recognition of their effort in the research, and V_{max} is the maximum reaction rate. Eq. (6.17) is the classical form of the kinetic equation of Michaelis–Menten.

6.3.1.3 Importance of V_{max} and K_m

The maximum reaction rate V_{max} represents the efficiency of the operation, in the upper limit, of a certain amount of an enzyme. When $r = V_{max}$, all the active sites are occupied and there are no free molecules of enzyme *E*. This condition is called *saturation at 100%*. When the saturation is 50%, that is $r = V_{max}/2$, the Michaelis–Menten equations establish that $K_m = [S]$. In this way, K_m (with units of concentration) represents the amount of substrate that is needed to fix half of the available enzyme and to produce half of the maximum reaction rate.

6.3.2 Methods to Determine the Rate Coefficients of the Michaelis–Menten Equation

6.3.2.1 Linear Regression

The following are approaches reported in the literature to obtain linear transformations of the Michaelis–Menten equation.

6.3.2.1.1   Lineweaver–Burk Method

Eq. (6.17) can be transformed in a linear equation by applying the inverse of both terms:

$$\frac{1}{r} = \frac{1}{V_{\max}} + \frac{K_m}{V_{\max}}\frac{1}{[S]} \tag{6.20}$$

6.3.2.1.2   Eadie–Hofstee Method

Eq. (6.17) can be reordered by dividing the numerator and denominator of the right-hand side by K_m and then multiplying both terms by $1/[S]$:

$$\frac{r}{[S]} = \frac{V_{\max}}{K_m} - \left(\frac{1}{K_m}\right)r \tag{6.21}$$

6.3.2.1.3   Augustinsson Method

Divide the numerator and denominator of the right-hand side of Eq. (6.17) by $[S]$, and reorder:

$$r = V_{\max} - K_m\frac{r}{[S]} \tag{6.22}$$

6.3.2.1.4   Woolf Method

From the final equation of Lineweaver and Burk (1934), both terms are multiplied by $[S]$:

$$\frac{[S]}{r} = \frac{K_m}{V_{\max}} + \frac{1}{V_{\max}}[S] \tag{6.23}$$

6.3.2.2   Graphic Method

A plot of $[S]$ versus v (Figure 6.16) allows for direct calculation of V_{max}, and from its value, K_m can be calculated. The calculation of K_m is simple, because $K_m = [S]$ for $r = V_{\max}/2$.

6.3.2.3   Integral Method

K_m and V_{max} can be calculated from experimental data of time (t) and concentration of the substrate (C_s), starting from Eq. (6.17):

$$r = -\frac{dCs}{dt} = \frac{V_{\max}Cs}{K_m + Cs} \tag{6.24}$$

Separating variables, integrating and rearranging:

$$\frac{\ln^{Cs_o}/_{Cs}}{Cs_o - Cs} = \frac{V_{\max}}{K_m}\left(\frac{t}{Cs_0 - Cs}\right) - \frac{1}{K_m} \tag{6.25}$$

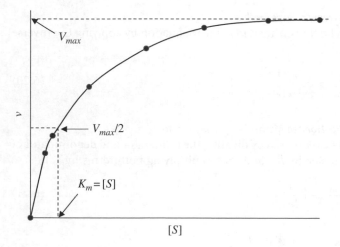

Figure 6.16 Graphical representation of the Michaelis–Menten equation.

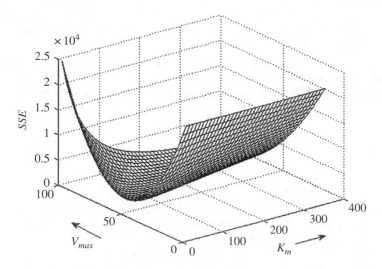

Figure 6.17 Optimal values of parameters of the Michaelis–Menten equation.

6.3.2.4 Non-linear Regression

An additional way to calculate V_{max} and K_m is by means of non-lineal regression analysis. The most used method is that proposed by Mar-quardt–Levenberg, which is based on the minimization of an objective function that is generally the *SSE* between experimental and calculated data. Figure 6.17 shows a typical representation of the objective function, in which the minimum value is observed:

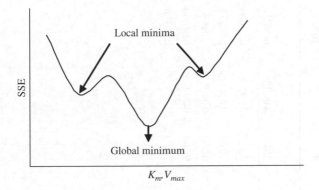

Figure 6.18 Global minimum and local minima in a non-lineal regression.

$$SSE = \sum \left| \left(r^{calc} - r^{exp} \right) \right|^2 = \sum_{i=1}^{N} \left| \frac{V_{max}\, C_s^{exp}}{K_m + C_s^{exp}} - r^{exp} \right| \tag{6.26}$$

To begin with the non-linear regression analysis, the initial values of the unknown parameters need to be assumed. In most cases, these assumptions do not need to be exact. However, if they are far from the optimum values, the minimization algorithm can converge to local minima and not to the global minimum (Figure 6.18). To assure that the estimated parameters correspond to the global minimum, the sensitivity analysis described in Section 6.2.2.4 can be used.

6.3.3 Application of the Methods

6.3.3.1 Experimental Data
The following ten cases of experimental data, reported in several sources (Table 6.9) for enzymatic reactions, were used to compare the calculation of Michaelis–Menten parameters with the different methods previously described:

1) Glutamic acid dehydrogenase enzyme
2) Phosphoenolpyruvate (*PEP*) in the presence of pyruvate kinase
3) Reaction catalysed by enzymes
4) Acid phosphatase enzyme
5) Hydrolysis of methyl hydrocinnamate in the presence of chymotrypsin enzyme
6) Oxidation of sodium succinate to fumarate in the presence of sodium succinate dehydrogenase enzyme
7) Hydrolysis of the ester of N-acetyl-phenylalanine ρ-nitrophenol and α-chymotrypsin with bovine pancreas

Table 6.9 Experimental data reported in the literature for enzymatic reactions (taken from different sources).

Data set	1	2	3	4	5	6	7	8	9	10
					Concentration of substrate, [S]					
Units	µM	µM	M	µM	M dm^3	M dm^{-3}	µM	µM	Kmol/m^3	mol/L
	1.68	0.020	0.5×10^{-4}	0.50	0.32×10^{-3}	0.33×10^{-3}	10	0.125	0.002	0.02000
	3.33	0.030	1.0×10^{-4}	0.75	1.28×10^{-3}	0.50×10^{-3}	15	0.167	0.010	0.01775
	5.00	0.055	2.0×10^{-4}	2.00	2.24×10^{-3}	1.00×10^{-3}	20	0.250	0.020	0.01580
	6.67	0.085	3.5×10^{-4}	4.00	4.60×10^{-3}	2.00×10^{-3}	25	0.500	0.200	0.01060
	10.00	0.150	5.0×10^{-4}	6.00	8.57×10^{-3}	10.2×10^{-3}	30	1.000		0.00500
	20.00	0.200		8.20	14.6×10^{-3}		35			
				10.00	30.8×10^{-3}		40			
							45			
							50			

					Reaction rate, r					
Units	ΔA_{360}/min	µM/min	mM/min	ΔA_{405}/min	10^8/M dm^{-3} s^{-1}	µM/s	µM/min	Min	Kmol/m^3 s	Min[a]
	0.172	0.090	0.71	0.075	1.5	0.50	11.7	25.0	0.09	0
	0.250	0.104	1.07	0.090	5.0	0.62	15.0	30.3	0.20	10
	0.286	0.115	1.50	0.152	7.5	0.79	17.5	40.4	0.38	20
	0.303	0.118	1.80	0.196	11.5	0.99	19.4	55.6	0.55	50
	0.334	0.124	1.88	0.210	15.0	1.17	21.0	71.4	1.08	100
		0.130		0.214	17.5		2.3			
				0.230	20.0		23.3			
							24.2			
							25.0			

a) These data series are reported in time.

8) Reaction catalysed by enzymes
9) Hydrolysis of urea in the presence of the enzyme urease
10) Concentration of hydrogen peroxide versus time.

6.3.3.2 Calculation of Kinetic Parameters

V_{max} and K_m were calculated for all the cases commented on before. As an example, Figure 6.19 shows the fits obtained for Case 1 with the four linear transformations of the equation of Michaelis–Menten.

Figure 6.20 shows the application of the graphical method for the same case, whereby V_{max} was first calculated. To determine K_m, the experimental data are adjusted by linear regression to the following equation: $r = a \ln^0 C_S + b$. Because $K_m = C_s$ for $r = V_{max}/2$, K_m is calculated with:

$$K_m = EXP \left[\frac{\frac{V_{max}}{2} - b}{a} \right] \tag{6.27}$$

In the case of non-lineal regression analysis, the method of Marquardt–Levenberg is available in commercial software such as Matlab, by means of which the corresponding calculations were done.

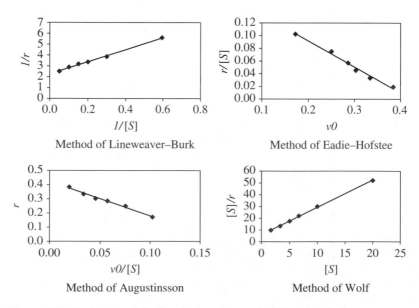

Method of Lineweaver–Burk

Method of Eadie–Hofstee

Method of Augustinsson

Method of Wolf

Figure 6.19 Application of the lineal regression method for example 1.

For the case in which experimental data of time versus concentration of substrate is reported, to apply the different methods, it is necessary to find a relationship between C_s and t. The best fit was the following polynomial:

$$C_s = 1.9977 \times 10^{-2} + 2.2535 \times 10^{-4}t + 7.5779 \times 10^{-7}t^2 \qquad (6.28)$$

This equation was derived with respect to time, and the following expression was obtained, whereby reaction rates were calculated:

$$r = -\frac{dC_s}{dt} = 1.9977 \times 10^{-2} + 4.507 \times 10^{-4}t \qquad (6.29)$$

Once the values of the reaction rates are determined, the calculation of V_{max} and K_m was done by means of the different methods.

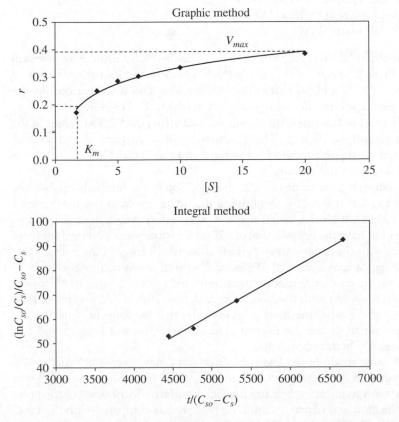

Figure 6.20 Representation of the graphical method (Case 1) and integral method.

The use of the integral method for this case is illustrated in Figure 6.20. The linear regression analysis with this method gives the following values:

$$K_m = 3.4413 \times 10^{-2}$$

$$V_{max} = 6.2577 \times 10^{-4}$$

$$r^2 = 0.9961$$

$$SSE = 6.49604 \times 10^{-11}$$

The results for the other cases and methods are detailed in Table 6.10.

6.3.4 Discussion of Results

The results of the correlation coefficient (r^2) obtained with the different methods are presented in Figure 6.21. The four methods were named as follows:

- Lineweaver–Burk method (L-B)
- Eadie–Hofstee method (E-H)
- Augustinsson method (A)
- Woolf method (W).

It is observed from Figure 6.21 that the Eadie–Hofstee and Augustinsson methods give equal values of r^2, which is because the former uses a plot of r versus $r/[S]$ and the latter uses $r/[S]$ versus r. This is the reason why the Augustinsson method is typically attributed to Eadie–Hofstee.

In most of the cases, the Woolf method yields the highest values of the correlation coefficient. The Lineweaver–Burk method is the second, the Augustinsson method is the third and the Eadie–Hofstee method is the fourth in the ranking.

Comparing the values of *SSE*, the precision of the methods exhibits the same order. It is worth mentioning that there are cases (such as Cases 1 and 2) in which the Woolf method, although having a high value of r^2, does not have the lowest value of *SSE*. The contrary was observed for these cases with the Lineweaver–Burk method, which showed the lowest value of r^2 but a lower value of *SSE* compared with the Woolf method.

In some cases, the four methods reported the same value of r^2 close to unity; however, with the Woolf method, the value of *SSE* was lower than those of the other methods. This indicates that the value of r^2 itself is not adequate to define the best fit. Thus, examination of both r^2 and *SSE* values is a better approach.

A comparison of the Eadie–Hofstee and Lineweaver–Burk methods is reported in the literature, which indicated that the former may hide lower deviation to linearity, because it yields a uniform distribution of the point in the plot, and hence it should be preferred as compared with the Lineweaver–Burk method, which agrees with the calculations done with the ten cases in this section. However, when comparing the four methods,

Table 6.10 Results of the estimation kinetic parameters.

	Method	Lineweaver and Burk	Eadie and Hofstee	Augustinsson	Woolf	Graphic	Non-linear regression
Case 1	V_{max}	4.2288×10^{-1}	4.2354×10^{-1}	4.2196×10^{-1}	4.3095×10^{-1}	3.900×10^{-1}	4.2380×10^{-1}
	K_m	2.4306	2.4415	2.4130	2.6064	1.9009	2.4519
	r^2	0.9968	0.9883	0.9883	0.9990	–	–
	SSE	1.6825×10^{-4}	5.5720×10^{-3}	3.0894×10^{-4}	2.0143×10^{-4}	–	1.6757×10^{-4}
Case 2	V_{max}	1.3372×10^{-1}	1.3402×10^{-1}	1.3349×10^{-1}	1.3494×10^{-1}	1.3000×10^{-1}	1.3350×10^{-1}
	K_m	9.4002×10^{-3}	9.5318×10^{-3}	9.2809×10^{-3}	1.0245×10^{-2}	2.9827×10^{-3}	9.2629×10^{-3}
	r^2	0.9843	0.9736	0.9736	0.9991	–	–
	SSE	2.0693×10^{-5}	3.9555×10^{-3}	2.7580×10^{-5}	2.7697×10^{-5}	–	2.0556×10^{-5}
Case 3	V_{max}	2.3491	2.3623	2.3571	2.3323	1.9000	2.3526
	K_m	1.1603×10^{-4}	1.1746×10^{-4}	1.1684×10^{-4}	1.1340×10^{-4}	7.6944×10^{-5}	1.1606×10^{-4}
	r^2	0.9991	0.9947	0.9947	0.9987	–	–
	SSE	2.4774×10^{-3}	4.7392×10^{-2}	5.2209×10^{-3}	2.5980×10^{-3}	–	2.4558×10^{-3}
Case 4	V_{max}	2.4834×10^{-1}	2.5388×10^{-1}	2.5245×10^{-1}	2.5550×10^{-1}	2.3000×10^{-1}	2.5530×10^{-1}
	K_m	1.2063	1.2735	1.2527	1.3082	1.0666	1.3073
	r^2	0.9922	0.9836	0.9836	0.9982	–	–
	SSE	1.5459×10^{-4}	3.9778×10^{-3}	3.8375×10^{-4}	1.0345×10^{-4}	–	1.0337×10^{-4}
Case 5	V_{max}	2.2868×10^{-7}	2.3002×10^{-7}	2.3007×10^{-7}	2.2992×10^{-7}	2.0000×10^{-7}	2.3013×10^{-7}
	K_m	4.5597×10^{-3}	4.6019×10^{-3}	4.6017×10^{-3}	4.5959×10^{-3}	3.0712×10^{-3}	4.6067×10^{-3}
	r^2	1.0000	0.9999	0.9999	1.0000	–	–
	SSE	1.9140×10^{-8}	2.2817×10^{-18}	1.7454×10^{-18}	2.6015×10^{-9}	–	0

(Continued)

Table 6.10 (Continued)

	Method	Lineweaver and Burk	Eadie and Hofstee	Augustinsson	Woolf	Graphic	Non-linear regression
Case 6	V_{max}	1.2009	1.2170	1.2136	1.2275	1.2100	1.2221
	K_m	4.6748×10^{-4}	4.8495×10^{-4}	4.8086×10^{-4}	5.0094×10^{-4}	4.3975×10^{-4}	4.9544×10^{-4}
	r^2	0.9969	0.9915	0.9915	0.9999	–	–
	SSE	1.5599×10^{-3}	4.4343×10^{-2}	2.4845×10^{-3}	1.0470×10^{-3}	–	1.0402×10^{-3}
Case 7	V_{max}	34.859	34.900	34.898	34.931	25.000	34.9329
	K_m	19.819	19.871	19.869	19.915	10.988	19.920
	r^2	0.9999	0.9999	0.9999	0.9999	–	–
	SSE	5.8339×10^{-3}	3.4690×10^{-2}	1.5684×10^{-2}	4.8066×10^{-3}	–	4.7977×10^{-3}
Case 8	V_{max}	96.365	96.750	96.624	97.228	72.000	97.0346
	K_m	3.5865×10^{-1}	3.6114×10^{-1}	3.6027×10^{-1}	3.6478×10^{-1}	2.0908×10^{-1}	3.6361×10^{-1}
	r^2	0.9994	0.9975	0.9975	0.9996	–	–
	SSE	8.4538×10^{-1}	4.6268	3.4908	7.7215×10^{-1}	–	6.9889×10^{-1}
Case 9	V_{max}	1.3279	1.2353	1.2216	1.2101	1.1000	1.2058
	K_m	2.7523×10^{-2}	2.4865×10^{-2}	2.4427×10^{-2}	2.4014×10^{-2}	1.9794×10^{-2}	2.3332×10^{-2}
	r^2	0.9992	0.9823	0.9823	0.9997	–	–
	SSE	8.3898×10^{-3}	2.9959×10^{-2}	1.0629×10^{-2}	6.7014×10^{-4}	–	5.6653×10^{-4}
Case 10	V_{max}	7.8604×10^{-4}	7.0836×10^{-4}	6.6315×10^{-4}	6.7965×10^{-4}	2.2500×10^{-3}	6.7997×10^{-4}
	K_m	4.7629×10^{-2}	4.1707×10^{-2}	3.8199×10^{-2}	3.9499×10^{-2}	7.2527×10^{-3}	3.9497×10^{-2}
	r^2	0.9775	0.9158	0.9158	0.9576	–	–
	SSE	1.0588×10^{-10}	1.0715×10^{-10}	1.2532×10^{-9}	5.1130×10^{-11}	–	5.1760×10^{-11}

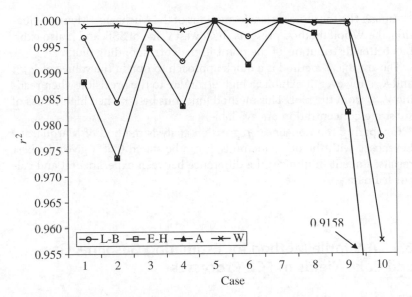

Figure 6.21 Comparison of R^2 for the four methods of lineal regression.

it was found that better distribution of the point in the plot was achieved with the Woolf method, as observed in Figure 6.19.

From the previous comparisons, it can be clearly seen that from the four linear transformations of the Michaelis-Menten equation, the Woolf method is the best one.

Comparing the results obtained with the four methods and with the integral method, it is observed that the value of r^2 with the latter is higher ($r^2 = 0.9961$) than that obtained with the other methods, although the value of *SSE* with the Woolf method is still the lowest.

Comparing the values of *SSE* obtained with the Woolf method and those with the non-linear regression analysis with the Marquardt–Levenberg method, it is seen that, in general, the latter gives better results (lower values of *SSE*).

6.3.5 Conclusions

The following are the four linear transformations of the Michaelis–Menten equation reported in the literature: Lineweaver–Burk, Eadie–Hofstee, Augustinsson and Woolf.

The correlation coefficient itself is not adequate to define which model gives the best fit. It has been shown that the combination of r^2 and SSE is a better approach.

Among the four linear transformations of the Michaelis–Menten equation, the Woolf method presents the lowest value of SSE, and it also exhibits better distribution of the point in the corresponding plot.

The graphical method is a quick approach to obtain the values of V_{max} and K_m; however, it exhibits a high error due to the precision when reading V_{max} from the plot. This method functions better when high values of substrate concentration are available.

Comparing the non-linear regression analysis method of Marquardt–Levenberg with the other methods, it can be stated that it gives the best results, since it minimizes the difference between experimental and calculated data.

6.4 A Simple Method for Estimating Gasoline, Gas and Coke Yields in FCC Processes

6.4.1 Introduction

Fluid catalytic cracking (FCC) is one of the most important conversion processes in a petroleum refinery. The objective of the FCC unit is to convert low-value, high-boiling-point feedstocks into more valuable products such as gasoline and lighter products.

Cracking reactions also produce deposition of coke on the catalyst, and catalyst activity is restored by burning off the coke with air in the regenerator. The catalyst-burning step supplies the heat for the reactions through circulation of catalyst between reactor and regenerator (Sadeghbeigi, 1995).

Gasoline formation is of major importance for the development of the FCC catalyst because it determines the selectivity of the desired and unwanted products.

The MAT (Microactivity Test) technique, a normalized ASTM procedure that uses a small tubular reactor for a standard feedstock which allows one to change easily the reaction conditions, is the one of the most common ways for catalyst evaluation and kinetic measurements.

Most of the kinetic models available in the literature predict the product yields using only a few lumps. This is because models with many pseudo-components require more experimental information in order to estimate their kinetic parameters, and they cannot be assumed to be more precise than simpler models. The only difference is that higher parameter models predict the products distribution with more detail. However, the differential equations derived from reaction kinetics and reactor mass balance are not applicable for the evaluation of routine catalyst screening because of the small number of observations available compared with the number of estimated parameters.

In this section, a simple method for calculating gasoline, gas and coke yields–conversion relationships in the FCC process is evaluated. The kinetic calculations are done using experimental information obtained in a MAT unit with catalyst and feedstock recovered from an industrial unit, and experimental information reported in the literature.

6.4.2 Methodology

6.4.2.1 Choosing the Kinetic Models

Many complex reactions occur in the catalytic cracking process, but the ones of primary interest are those that crack molecules into smaller ones and thus reduce their boiling point to the more useful range of gasoline and light products (Krambeck, 1991).

The first method to obtain a kinetic representation of complex reactions was to lump molecules in distillation cuts and to consider pseudo-chemical reactions between these lumps.

In the case of gas oil and heavier fractions that are fed into FCC reactors, they can contain thousands of different compounds, producing many different products. The earliest kinetic study of this process focused on the conversion of heavy oil to gasoline and was based on just two lumps: those materials boiling above the gasoline range and everything else (Blanding, 1953).

At the other extreme is the single-events method, the most advanced method (Feng *et al.*, 1993) that permits a mechanistic description of catalytic cracking based on detailed knowledge of the mechanism of the various reactions involving carbenium ions. However, the application of this method to catalytic cracking of real feedstocks is difficult because of analytical complexity and computational limitations (Van Landeghem *et al.*, 1994). Moreover, for reactor design and simulation purposes, kinetic models with a few lumps have been found to describe the FCC process with acceptable accuracy; these models take into account the major industrial products in the catalytic cracking process.

On the other hand, the more lumps a model includes, intrinsically more kinetic parameters will need to be estimated, and, consequently, more experimental information is required.

Based on the aforementioned considerations and taking into account that coke formation (which supplies the heat required for the heating and vaporization of the feedstock and to perform the endothermic reactions) and gas production become very important to design and simulate the air blower and gas compressor, respectively, a four-lump kinetic model was used to evaluate gasoline, gas and coke yields.

6.4.2.2 Reaction Kinetics

In order to evaluate the kinetic parameters included in the four-lump model, it is easier to calculate first some parameters with a three-lump model that will be the same in both models (e.g. a gas oil to gasoline cracking kinetic constant) (Ancheyta *et al.*, 1997). The three-lump model (Figure 6.22) involves parallel cracking of gas oil (y_1) to gasoline (y_2) and gas plus coke (y_3), with consecutive cracking of the gasoline to gas plus coke (Weekman, 1968). The four-lump model (Figure 6.23) involves also parallel cracking of gas oil (y_1) to gasoline (y_2), gas (y_{31}) and coke (y_{32}), with consecutive cracking of the gasoline to gas and coke (Lee *et al.*, 1989).

For gas oil cracking the rate is assumed to be second order, and for gasoline first order. The kinetic rate equations for these models are as follows (*t*: time, k_i: kinetic constant). ϕ, the catalyst decay function, was considered equal for all reactions; this means that a non-selective deactivation function was used.

Three-lump kinetic model:

$$\frac{dy_1}{dt} = -k_1 y_1^2 \phi - k_3 y_1^2 \phi = -(k_1 + k_3) y_1^2 \phi = -k_o y_1^2 \phi \tag{6.30}$$

$$\frac{dy_2}{dt} = k_1 y_1^2 \phi - k_2 y_2 \phi = \left(k_1 y_1^2 - k_2 y_2\right)\phi \tag{6.31}$$

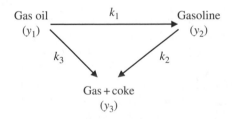

Figure 6.22 Three-lump kinetic model.

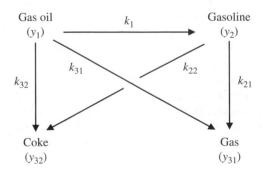

Figure 6.23 Four-lump kinetic model.

$$\frac{dy_3}{dt} = \left(k_3 y_1^2 + k_2 y_2\right)\phi \tag{6.32}$$

Four-lump kinetic model:

$$\frac{dy_1}{dt} = -k_1 y_1^2 \phi - k_{31} y_1^2 \phi - k_{32} y_1^2 \phi = -(k_1 + k_3) y_1^2 \phi = -k_o y_1^2 \phi \tag{6.33}$$

$$\frac{dy_2}{dt} = k_1 y_1^2 \phi - k_{21} y_2 \phi - k_{22} y_2 \phi = \left(k_1 y_1^2 - k_2 y_2\right)\phi \tag{6.34}$$

$$\frac{dy_{31}}{dt} = \left(k_{31} y_1^2 + k_{21} y_2\right)\phi \tag{6.35}$$

$$\frac{dy_{32}}{dt} = \left(k_{32} y_1^2 + k_{22} y_2\right)\phi \tag{6.36}$$

6.4.2.3 Estimation of Kinetic Parameters

The kinetic parameters can be estimated using Eqs. (6.30) to (6.36). Firstly, the variation with time-of-yields experimental data (gas oil, gasoline, gas and coke) can be fitted using the following polynomial functions (a_i: polynomial constants, n: polynomial order, y_i: product yield and y_i': time derivatives).

$$y_i = a_{i,0} + a_{i,1}t + a_{i,2}t^2 + \ldots + a_{i,n}t^n \tag{6.37}$$

$$y_i' = \frac{dy_i}{dt} = a_{i,1} + 2a_{i,2}t + \ldots + na_{i,n}t^{n-1} \tag{6.38}$$

Combining Eqs. (6.30) to (6.36), the following expressions can be obtained, which are straight lines:

$$-\left(\frac{y_1' + y_2'}{y_2}\right) = k_2 \phi + k_3 \phi \left(\frac{y_1^2}{y_2}\right) \tag{6.39}$$

$$\left(\frac{y_1' + y_3'}{y_2}\right) = k_2 \phi - k_1 \phi \left(\frac{y_1^2}{y_2}\right) \tag{6.40}$$

$$-\left(\frac{y_1' + y_{31}'}{y_2}\right) = -k_{21} \phi + (k_1 \phi + k_{32} \phi)\left(\frac{y_1^2}{y_2}\right) \tag{6.41}$$

The left-hand side (LHS) of Eqs. (6.39)–(6.41) can be evaluated from Eqs. (6.37) and (6.38), which were determined with experimental data. With the intercept and slope of Eqs. (6.39)–(6.41), $k_1\phi$, $k_2\phi$, $k_3\phi$, $k_{21}\phi$ and $k_{32}\phi$ could be evaluated. The other parameters ($k_{22}\phi$ and $k_{31}\phi$) can be obtained using the following equations:

$$k_2 \phi = k_{21} \phi + k_{22} \phi \tag{6.42}$$

$$k_3 \phi = k_{31} \phi + k_{32} \phi \tag{6.43}$$

6.4.2.4 Evaluation of Products Yields

Gasoline, gas and coke yields–conversion relationships can be obtained by dividing Eqs. (6.34), (6.35) and (6.36) by (6.30). The resulting equations are the following ($x = 1-y_1$ is the conversion):

$$\frac{dy_2}{dx} = \frac{k_1\phi}{k_0\phi} - \frac{k_2\phi}{k_0\phi}\left(\frac{y_2}{y_1^2}\right) = r_1 - r_2\left(\frac{y_2}{y_1^2}\right) \tag{6.44}$$

$$\frac{dy_{31}}{dx} = \frac{k_{31}\phi}{k_0\phi} + \frac{k_{21}\phi}{k_0\phi}\left(\frac{y_2}{y_1^2}\right) = r_{31} + r_{21}\left(\frac{y_2}{y_1^2}\right) \tag{6.45}$$

$$\frac{dy_{32}}{dx} = \frac{k_{32}\phi}{k_0\phi} + \frac{k_{22}\phi}{k_0\phi}\left(\frac{y_2}{y_1^2}\right) = r_{32} + r_{22}\left(\frac{y_2}{y_1^2}\right) \tag{6.46}$$

where:

$$r_1 = \frac{k_1\phi}{k_0\phi} = \frac{k_1}{k_0} \tag{6.47}$$

$$r_2 = \frac{k_2\phi}{k_0\phi} = \frac{k_2}{k_0} \tag{6.48}$$

$$r_{21} = \frac{k_{21}\phi}{k_0\phi} = \frac{k_{21}}{k_0} \tag{6.49}$$

$$r_{22} = \frac{k_{22}\phi}{k_0\phi} = \frac{k_{22}}{k_0} \tag{6.50}$$

$$r_{31} = \frac{k_{31}\phi}{k_0\phi} = \frac{k_{31}}{k_0} \tag{6.51}$$

$$r_{32} = \frac{k_{32}\phi}{k_0\phi} = \frac{k_{32}}{k_0} \tag{6.52}$$

6.4.2.5 Advantages and Limitations of the Methodology

This methodology can be used to evaluate the combined kinetic parameters and decay function ($k_i\phi$) involved in the four-lump kinetic model using experimental data obtained in a micro-activity plant and a simple linear regression analysis. With this method, the initializing problems of kinetic parameters values that could converge to a local minimum of the objective function, which are frequently found in non-linear parameter estimation, could be avoided.

The kinetic models based on lumping methodology for catalytic cracking reactions, in which some of the products are lumped and treated kinetically as one species with various cracking reaction orders, have been widely used in the most advanced riser models. However, the weakness of these models is that the kinetic constants are a function of feedstock and catalyst properties.

The values of the kinetic parameters obtained with this method can be used as initial values to estimate kinetic constants in models with more than four lumps [i.e. five-lump (Ancheyta *et al.*, 1999) and six-lump (Corella and Frances, 1991)], and the convergence to a local minimum of the objective function may be reduced.

6.4.3 Results and Discussion

A summary of the experimental information reported by Wang (1970) at a reaction temperature of 548.9 °C and a catalyst-to-oil ratio (C/O) of 4 was used firstly for validating the proposed method. The space–velocity was assumed to be the inverse of the reaction time. The variation of total conversion and the gasoline, gas and coke yields with time are shown in Table 6.11. These literature data were used to obtain the following polynomial functions between time and product yields. The best fit was found with a fifth-order polynomial (FOP) obtaining correlation coefficients very close to unity ($r^2 > 0.9999$).

$$y_1 = 1.0 - 0.78914t + 0.34639t^2 - 4.37482 \times 10^{-2}t^3$$
$$- 7.61929 \times 10^{-3}t^4 + 1.38442 \times 10^{-3}t^5 \tag{6.53}$$

$$y_2 = 0.64023t - 0.31049t^2 + 0.04167t^3 + 6.67416 \times 10^{-3}t^4$$
$$- 1.25674 \times 10^{-3}t^5 \tag{6.54}$$

$$y_3 = 0.15132t - 4.06585 \times 10^{-2}t^2 + 5.15936 \times 10^{-3}t^3$$
$$+ 1.63206 \times 10^{-4}t^4 - 6.27515 \times 10^{-5}t^5 \tag{6.55}$$

$$y_{31} = 0.11722t - 3.20337 \times 10^{-2}t^2 + 2.70731 \times 10^{-3}t^3$$
$$+ 7.22615 \times 10^{-4}t^4 - 1.10677 \times 10^{-4}t^5 \tag{6.56}$$

Table 6.11 Summary of experimental data reported in the literature at 548.9 °C and C/O of 4 (taken from Wang, 1970).

Space velocity (hr^{-1})	Conversion (wt%)	Gas oil yield (wt%)	Gasoline yield (wt%)	Gas yield (wt%)	Coke yield (wt%)
10	82.38	17.62	54.16	21.08	7.14
20	71.18	28.82	48.65	16.81	5.72
30	62.04	37.96	43.85	13.60	4.59
60	49.26	50.74	37.67	8.85	2.74

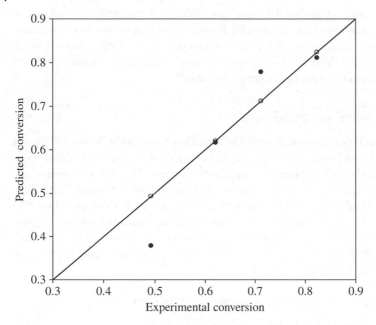

Figure 6.24 Comparison between SOP (●) and FOP (o) for conversion prediction.

Figure 6.24 shows a comparison between experimental (Table 6.11) and calculated conversions using Eq. (6.53) ($x = 1 - y_1$, where x is conversion, and y_1 is unconverted gas oil) and those determined with a second-order polynomial (SOP), as were used by Wallenstein and Alkemade (1996) to compare their OPE functions (optimum performance envelope). It is clear that FOP is more accurate than SOP. The reduction in absolute error is better than that reported in reference 2, where the SOP was compared with OPE. This means that Eqs. (6.53)–(6.56) are adequate to represent the variation with time of FCC product yields.

The corresponding time derivatives of Eqs. (6.25)–(6.28) are:

$$y_1' = -0.78914 + 0.69278t - 0.13124t^2 - 0.03048t^3 + 0.00692t^4$$
$$(6.57)$$

$$y_2' = 0.64023 - 0.62098t + 0.12501t^2 + 0.02669t^3 - 0.00628t^4$$
$$(6.58)$$

$$y_3' = 0.15132 - 0.08132t + 0.01547t^2 + 0.00065t^3 - 0.00031t^4$$
$$(6.59)$$

$$y_{31}' = 0.11722 - 0.06407t + 0.00812t^2 + 0.00029t^3 - 0.00055t^4$$
$$(6.60)$$

The LHS of the straight lines given by Eqs. (6.39)–(6.41) can be obtained by using the polynomial [Eqs. (6.53)–(6.56)] and time derivatives expressions [Eqs. (6.57)–(6.60)]. With these expressions, more points than those determined experimentally can be calculated.

The kinetic constants obtained with intercept and slope lines are: $k_1\phi = 0.7116$, $k_2\phi = 0.0438$, $k_3\phi = 0.2458$, $k_{21}\phi = 0.0354$ and $k_{32}\phi = 0.0544$. The other kinetic constants calculated with Eqs. (6.42)–(6.43) are: $k_{22}\phi = 0.0084$ and $k_{31}\phi = 0.1914$. These calculated parameters are combined cracking and decay constants because they include the catalyst deactivation function (ϕ).

By using the combined cracking and decay constants $(k_i\phi)$ and Eqs. (6.47) to (6.52), the values of $r_1, r_2, r_{21}, r_{22}, r_{31}$ and r_{32} can be calculated, which are 0.7433, 0.0458, 0.0369, 0.0088, 0.1999 and 0.0568, respectively.

With these values, Eqs. (6.44)–(6.46) could be solved numerically using a fourth-order Runge–Kutta method with the boundary condition $y_1 = 1$, $y_2 = y_{31} = y_{32} = 0$ at $x = 0$.

Figure 6.25 shows the experimental and predicted yields for gasoline, gas and coke. It can be seen that the kinetic parameters obtained with the proposed methodology predicted very well the products yields, with average deviations less than 3%.

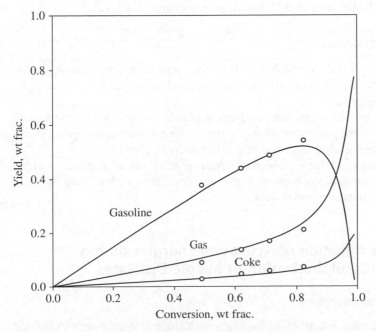

Figure 6.25 Experimental (symbols) and predicted (lines) gasoline, gas and coke yields using data reported in the literature.

Hari *et al.* (1995) followed a similar procedure to evaluated the combined cracking and decay constants ($k_i\varphi$) with experimental data obtained in a MAT unit at a reaction temperature of 528 °C, WHSV of 10 h^{-1} and catalyst-to-oil ratio of 3.65 using a modified three-lump kinetic model (gas oil: over 370 °C; gasoline and middle distillates: C_5–370 °C, and coke plus gas). Gasoline plus middle distillates formation and cracking reactions were assumed to be a first-order reaction, while gas oil to gas plus coke cracking was considered a second-order reaction.

They used an exponential law ($\phi = e^{-k_d t_c}$) taken from the literature (Weekman, 1969) to evaluate the deactivation constant (k_d) in order to calculate the individual kinetic parameters (k_1, k_2 and k_3). However, as was shown before, it is not necessary to evaluate the decay function (ϕ), because it can be eliminated by using the kinetic constants relationships given by Eqs. (6.47) to (6.52).

The use of a deactivation function obtained from a correlation published in the literature (Weekman, 1969) is probably one of the reasons because these authors obtained high deviations between experimental and predicted yield values (15–20%), since the experimental information was determined using different feedstocks, catalysts and operating conditions.

Another important reason that could explain the high deviations obtained by Hari *et al.* (1995) is that they assumed first order for gas oil to gasoline and middle distillates reactions.

6.4.4 Conclusions

An easy procedure which uses the three- and four-lump kinetic models has been used to evaluate gasoline, gas and coke yields in the catalytic cracking process.

The combined cracking and decay constants included in the kinetic models were estimated using a simple linear regression analysis and experimental data obtained in a micro-activity plant.

The methodology showed accurate predictions of products yields–conversion relationships with average deviations of less than 3% with respect to experimental data.

6.5 Estimation of Activation Energies during Hydrodesulphurization of Middle Distillates

6.5.1 Introduction

The complex nature of oil fractions with sulphur compounds exhibiting very different re-activities, as well as the presence of others such as nitrogen (basic and no-basic), aromatics and so on reacting at the same time and competing for the same active sites and also inhibiting effects of by-products of the same reactions (e.g. hydrogen sulphide), have limited hydrodesulphurization (HDS) experimental studies to model compounds ranging from easy-to-desulphurize (e.g. thiophene) to hard-to-desulphurize (e.g. 4,6-dimethyldibenzothiophene). These are the main reasons why few works have been reported dealing with experiments with real petroleum feedstocks under industrial conditions, since most of the times it is not simple to extract individual effects and one does not know which one to blame. However, when a catalyst formulation is almost ready for commercial application, experiments with real feeds are mandatory. Testing with real feeds, not only for exploring the commercial application of new catalyst formulations but also for process design and optimization studies, is a very important step for new technology development. For the latter issue, kinetic data obtained from experiments with real feeds are of great interest, since they are employed for reactor modelling, simulation and optimization.

Kinetic studies for hydrodesulphurization can be divided roughly into two parts: those conducted with model compounds (see Section 6.2), and those carried out with real feeds. In the case of HDS of real feeds, n^{th}-order kinetics with respect to total sulphur concentration is usually applied, in which n value depends on several factors, such as type and concentration of sulphur compounds, catalyst properties, type of feed, operating conditions and experimental system, among others. Some authors use their experimental information to calculate the reaction order, which commonly ranges from 1.5 to 2.5, and others prefer to assume the value of n and then calculate the kinetic constant. Most of the values of reaction orders increase as the sulphur content in the feed also increases. However, this general increasing tendency of n with respect to sulphur in feed was not observed for activation energies, which may be due to differences in experimental conditions.

To get more knowledge about the kinetics of HDS of real feeds, this section shows the results of systematic studies to evaluate the effect of the type of feed on the reaction order and activation energy using experimental data obtained in a batch autoclave with a commercial catalyst and different petroleum distillates.

6.5.2 Experiments

Four petroleum products were used for hydrodesulphurization tests. Three were straight-run distillates (QS: kerosene, LSRGO: light straight-run gas oil, and HSRGO: heavy straight-run gas oil), and one was a cracked product from an FCC unit (LCO: light cycle oil). The main

Table 6.12 Properties of the feeds.

	QS	LCO	LSRGO	HSRGO
Specific gravity 20/4 °C	0.8386	0.8610	0.8626	0.9007
Total sulphur (% wt)	1.047	1.326	1.436	1.831
Viscosity at 40 °C, cSt	2.90	3.79	5.61	19.55

properties of the four feeds are presented in Table 6.12. A NiMo commercial catalyst was employed, and its properties are: 119 m^2/g specific surface area, 0.26 cm^3/g pore volume, 0.8983 g/cm^3 bulk density, 88 Å mean pore diameter, 2.42 wt% Co and 10.73 wt% Mo.

A schematic diagram of the reactor is shown in Figure 1.13. After catalyst loading, the reactor was closed and pressurized with hydrogen to the desired value. Heating was started from room temperature to the required value. Stirring was started when the temperature reached the set point, and the time was noted at that point as the beginning of the reaction. Temperature, pressure and stirring are controlled automatically using a digital controller. A total of five reaction products were collected each hour.

The experiments were conducted at constant pressure of 54 kg/cm^2, varying the reaction temperature and time in the ranges of 340–360 °C and 1–5 h, respectively. For each feed at each temperature, product samples (2–3 mL) were taken at 1 h intervals during 5 h for sulphur analysis. Taking into account that five samples were taken during the reaction, the total volume withdrawal (10–15 mL, <2% of the total reaction volume) was negligible.

6.5.3 Results and Discussion

6.5.3.1 Experimental Results

Figure 6.26a shows a comparison of sulphur content in the products for the four feeds at a reaction temperature of 350 °C. The typical behaviour is observed: the higher the reaction temperature and time, the lower the product sulphur content.

The possible hydrocracking of the four feeds was monitored by analysing the specific gravity of the products. Figure 6.26b presents the results at 350 °C as a function of time, from which one can conclude that hydrocracking is minimal at the reaction conditions studied in this work. In other words, the molecular weight of the feeds remains almost without change.

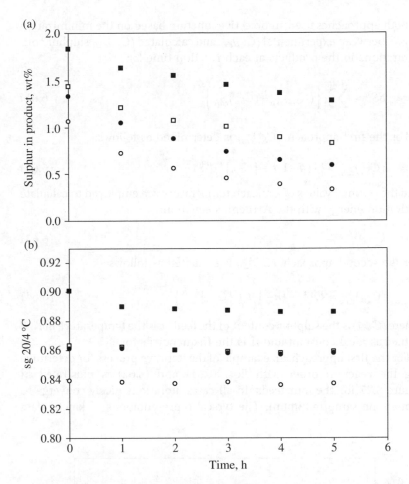

Figure 6.26 Effect of reaction temperature and time on (a) sulphur content in the product, (b) specific gravity of the product. (○) QS, (●) LCO, (□) LSRGO, (■) HSRGO.

6.5.3.2 Estimation of Kinetic Parameters

With data of sulphur content in hydrodesulphurized products at different times and temperatures, kinetic parameters were obtained by two approaches: (1) non-linear regression first to estimate the reaction order (n) and rate constant (k) as a function of temperature with the widely known integral method, and linear regression with the Arrhenius equation to determine activation energy (E_A); and (2) non-linear regression to calculate simultaneously n and E_A.

Both approaches used an objective function based on the minimization of *SSE* between experimental $(C_{S,i})_{exp}$ and calculated $(C_{S,i})_{calc}$ sulphur concentrations in the products at each reaction time (t):

$$SSE = \sum_{i=1}^{NData} \left\{ (C_{S,i})_{exp} - (C_{S,i})_{calc} \right\}^2 \tag{6.61}$$

For the first approach, $(C_{S,i})_{calc}$ is determined as follows:

$$(C_{S,i})_{calc} = \left\{ k(n-1)t + \{C_{S,o}\}^{1-n} \right\}^{\frac{1}{1-n}} \tag{6.62}$$

and the optimal values of k at each temperature are employed to calculate activation energy with the Arrhenius equation:

$$k = Ae^{-\frac{E_A}{RT}} \tag{6.63}$$

For the second approach, $(C_{S,i})_{cal}$ is evaluated as follows:

$$(C_{S,i})_{calc} = \left\{ Ae^{-\frac{E_A}{RT}}(n-1)t + \{C_{S,o}\}^{1-n} \right\}^{\frac{1}{1-n}} \tag{6.64}$$

where $(C_{S,i})_o$ is the sulphur content of the feed, T is the temperature in K, R is the gas ideal constant and A is the frequency factor.

For the first approach, an example of the iterative process for determining the reaction order with Eqs. (6.61) and (6.62) is illustrated in Figure 6.27 for the four feeds. In all cases, iterations nicely converge to almost the same minimum. The typical representations of kinetic data

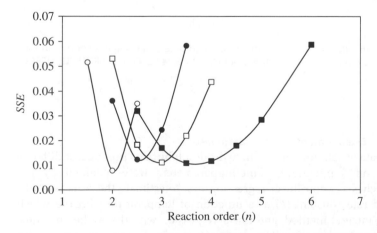

Figure 6.27 Sum of squared errors of the differences between experimental and calculated sulphur contents. (O) QS, (●) LCO, (□) LSRGO, (■) HSRGO.

with the integral method [i.e. $f(C_s)$ against time] and Arrhenius equation (Eq. 6.63) are plotted in Figures 6.28a and 6.28b. Correlation coefficients very close to or higher than 0.99 were obtained for all the cases. For the second approach, Eqs. (6.61) and (6.64) were solved, and the reaction order and activation energy for each feed were determined simultaneously. Kinetic parameters determined with both approaches are given in Table 6.13. It is observed from this table that in general, the first

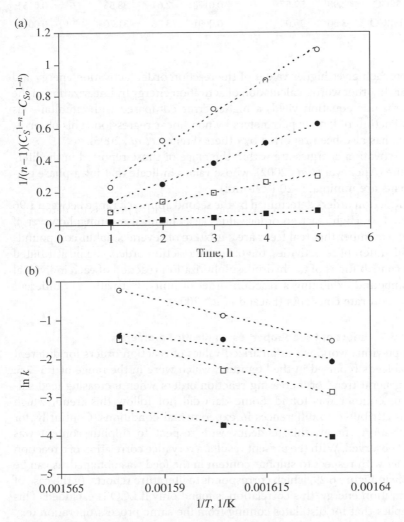

Figure 6.28 Linear representation of (a) power law model, (b) Arrhenius equation. (o) QS, (●) LCO, (□) LSRGO, (■) HSRGO.

Table 6.13 Comparison of kinetic parameter values determined with the two approaches.

	First approach			Second approach		
	n	E_A (Kcal/mol)	SSE	n	E_A (Kcal/mol)	SSE
QS	2.04	51.16	0.0085	1.96	41.96	0.0023
LCO	2.56	25.45	0.0284	2.32	21.49	0.0273
LSRGO	2.98	52.57	0.0178	2.61	38.55	0.0154
HSRGO	3.60	30.03	0.0091	3.36	31.70	0.0090

approach gives higher values of the reaction order, activation energy and *SSE*. In other words, calculation of activation energy by linearization of the Arrhenius equation yields a higher error compared with simultaneous estimation of kinetic parameters by non-linear regression. This observation has also been reported by others (Freitez *et al.*, 2005).

Activation energies are within the range of those reported in the literature (Ancheyta *et al.*, 2002), whose values indicate that intra-phase gradients are minimal (>20 Kcal/mol).

Reaction orders determined by the second approach range between 1.96 and 3.36. High reaction order values lack chemical meaning; however, if we remember that real feeds are a mixture of several sulphur compounds with different re-activities, the observed reaction orders can be attributed to contributions of the hydrodesulphurization reaction of each individual compound exhibiting a reaction order of unity, but with very different reaction rate constants (Bacaud *et al.*, 2002).

6.5.3.3 Effect of Feed Properties on Kinetic Parameters

In previous work, we summarized values of reaction orders for different real feeds reported in the literature, which were in the range of 1.5–2.5. A general trend of increasing reaction orders when increasing feed sulphur content was found. Some data did not follow this trend, which was attributed to differences in experimental conditions. Contrarily, for activation energy, this tendency with respect to sulphur content was not observed. With the present results, a very nice correlation of a reaction order with respect to sulphur content in the feed was obtained, as can be seen in Figure 6.29, which corresponds to literature reports. In the case of activation energy, this correlation is found only if LCO is excluded. This implies that for distillates coming from the same process/operation (e.g. crude oil atmospheric fractionation: QS, LSRGO and HSRGO), there is indeed a relationship between activation energy and sulphur content.

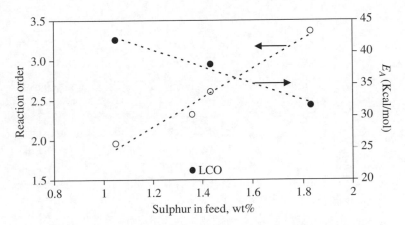

Figure 6.29 Relationship between reaction order (O) and activation energy (•) with sulphur content in the feed.

That is the reason why in previous literature, in which values of activation energies were taken from different literature sources corresponding to a variety of feeds also from different sources, this tendency was not observed.

6.5.4   Conclusions

Activation energies were determined in a batch autoclave for the hydro-desulphurization of kerosene, light straight-run gas oil, heavy straight-run gas oil and light cycle oil, which ranged from 21.49 to 41.96 Kcal/mol. Reaction kinetic orders were also determined in the range of 1.96 and 3.36. Although high values for reaction orders were obtained and probably without chemical meaning, the possible explanation of the observed values is that they correspond to the sum of contributions of the hydro-desulphurization reaction of each individual compound, which exhibit very different reactivity.

The reaction order showed a linear relationship with sulphur content in the feed, while activation energy showed that relationship only if cracked product (LCO) is excluded. This shows that the relationship between activation energy and feed properties depends on the source of petroleum fractions.

Problems

Irreversible Reactions of One Component

1) The decomposition of ammonia is carried out at 856 °C in a batch reactor at constant density, and the following data were obtained:

Time (sec)	0	200	400	600	1000
Total pressure (mmHg)	205	228	250	273	318

Determine:
A) The reaction order and the reaction rate coefficient
B) The half-life time
C) The pressure at the end of the reaction.

2) The following reaction in liquid phase is conducted at 36 °C, and the following data were obtained:

$$2A \rightarrow 3R + S$$

Time (h)	0	0.5	1.5	3	4	7	13
Concentration of A (mol/lt)	0.2	0.181	0.155	0.131	0.12	0.099	0.077

Determine:
A) The reaction order and the reaction rate coefficient
B) The time to react 90% conversion
C) If the activation energy is 10,800 cal/mol, calculate the half-life time at 60 °C.

Chemical Reaction Kinetics: Concepts, Methods and Case Studies, First Edition.
Jorge Ancheyta.
© 2017 John Wiley & Sons Ltd. Published 2017 by John Wiley & Sons Ltd.

3) If the reaction of problem 2 is written as:

$$A \rightarrow \frac{3}{2}R + \frac{1}{2}S$$

what would be the reaction order and the units of the reaction rate coefficient?

4) The following reaction in liquid phase:

$$A \rightarrow R + S$$

was studied in a batch reactor at constant density and temperature. The following kinetic model was obtained:

$$-r_A = 2.095 \times 10^{-2} \left(\text{min}^{-1}\right) C_A$$

Calculate the concentration of each component at 15 min.

5) For the following reaction in liquid phase:

$$A \rightarrow Products$$

the following experimental data were obtained in a batch reactor at 30 and 50 °C, by using an initial concentration of A of 1.5 mol/lt for both temperatures:

A) Find the reaction order and the reaction rate coefficient at 30 and 50 °C.

B) Calculate the temperature of the reactor to achieve 80% conversion in 50 min for $C_{A0} = 3$ mol/lt.

Temperature (°C)	Time (min)	Concentration of A (mol/lt)
	10	1.304
	20	1.154
30	40	0.937
	100	0.600
	10	1.003
	20	0.753
50	30	0.603
	40	0.503
	50	0.431

6) The hydrolysis of A in aqueous solution was carried out at 20 °C with an initial concentration of A of 0.05 mol/lt. The products of the hydrolysis ionize, so that they increase the electric resistance of the reaction mixture as the reaction proceeds. Note that the

electric conductivity is a lineal function of conversion. The measured values are:

Time (min)	0	5	10	20	40	80	150	200	240	∞
Electric resistance (ohm)	7.670	2.058	1.343	0.907	0.675	0.553	0.495	0.476	0.463	0.420

A) Find the reaction rate equation.

B) Calculate the reactor temperature to achieve 90% conversion in 100 min for the same initial concentration of A if the activation energy is 12,000 cal/mol.

7) For the decomposition of nitrous oxide at 895 °C, a value of the reaction rate coefficient of $k = 977$ $(mol/ml)^{-1}sec^{-1}$ was determined. If pure N_2O is heated at 1 atm and 895 °C in a batch reactor at constant density, calculate the required time to achieve 90% conversion.

8) The following experimental data correspond to the reaction in liquid phase at 50 °C:

$A \rightarrow Products$

Time (sec)	0	1800	3600	5400	9000	14400
C_A (mol/lt)	0.700	0.570	0.842	0.418	0.330	0.250

Calculate:

A) The reaction order and the reaction rate coefficient

B) The required time to reduce the concentration to a half of its initial value

C) If the activation energy is 10 Kcal/mol, how much time is needed to obtain 90% conversion at 100 °C?

9) The decomposition of phosphine was conducted in gas phase at 650 °C:

$4PH_3 \rightarrow P_4 + 6H_2$

and the reaction rate coefficient was found to be:

$$\log k = -\frac{18963}{T} + 2\log T + 12.13$$

where k has units of sec^{-1} and T of K. Assuming that the reaction is carried out in a closed vessel (constant density) at 1 atm of initial pressure, calculate the total pressure at 50, 100, 150 and 500 sec.

10) The following experimental data were reported for the decomposition of diazobenzene chloride in aqueous solution at 50 °C:

$$C_6H_5N_2Cl \rightarrow C_6H_5Cl + N_2$$

Time (min)	6	9	12	14	18	20	22	24	26	30
Released gas (ml)	19.30	26.00	32.60	36.00	41.30	43.30	45.00	46.50	48.50	50.35

After a long time period, a simple with an initial concentration of 10 g/lt released 58.3 ml of N_2. Determine the reaction order and the reaction rate coefficient.

11) 0.1 g of dimethyl ether and 0.05 g of nitrogen were loaded to a batch reactor at 504 °C and 1 atm. At these conditions, the reactant pyrolizes according to the reaction:

$$CH_3OCH_3 \rightarrow CH_4 + H_2 + CO$$

As the reaction proceeds, the following data of the time and volume differential were taken. Since pressure is constant, the variations in the reacting volume are registered by a movable plunger.

Time (sec)	2.5	5.3	8.5	12.2	16.6	21.8	28.8
$V - V_0$ (lt)	0.0275	0.0549	0.0824	0.1098	0.1373	0.1647	0.1922

Determine:

A) The reaction order and the reaction rate coefficient
B) The time to reach a volume that is double the initial value.

12) For the liquid phase reaction:

$$A \rightarrow R + S$$

The following value of the reaction rate coefficient was found at 180 °C for $C_{Ao} = 0.05$ mol/lt:

$$k = 0.134 \left(\frac{lt}{mol} \right)^{0.5} \text{min}^{-1}$$

Determine:
A) The half-life time
B) The initial concentration of A, if the half-life time is 1 h.

13) The following data were reported for the decomposition of pure nitrous oxide:

Effect of initial pressure of N_2O on $t_{1/2}$ at 1030 K				
Initial pressure of N_2O (mmHg)	82.5	139	296	360
Half-life time (sec)	860	470	255	212

Effect of initial pressure of N_2O and temperature on $t_{1/2}$			
Temperature (K)	1085	1030	907
Initial pressure of N_2O (mmHg)	345	360	294
Half-life time (sec)	53	212	1520

A) Find the reaction order and reaction rate coefficient.
B) If the reaction starts with 400 mmHg of N_2O, what is the reactor temperature to obtain 90% of conversion in 1000 sec?

14) For the reaction:

$$(C_6H_5)_3 - C - C - (C_6H_5)_3 \rightarrow 2(C_6H_5)_3 C^*$$

the following experimental data were obtained at $0\,°C$ in liquid phase (chloroform solution):

Time (sec)	0	35.4	174.0	313.0	759.0
C_A/C_{AO}	1.0	0.883	0.530	0.324	0.059

A) Find the reaction order and the reaction rate coefficient at $0\,°C$.
B) If the activation energy is 18,000 cal/mol, find the half-life time at $150\,°C$.

15) The reactant A is decomposed in solution at $100\,°C$ to form products. Starting with an initial concentration of 1.6 mol/lt, the following experimental data were obtained:

x_A	0.1	0.2	0.4	0.6	0.7
$-r_A$ (mol/lt min)	0.3190	0.2886	0.2260	0.1601	0.1254

The activation energy was found to be 8500 cal/mol. Calculate:
A) The time necessary to reach 75% of conversion at $100\,°C$
B) The reaction order
C) The reaction rate coefficient at $100\,°C$

D) The half-life time at 100 °C
E) The temperature for the half-life time to be half of the value observed at 100 °C.

16) The thermal decomposition in gas phase of reactant A:

$$A \rightarrow R + S + D$$

was carried out in a closed vessel at constant total pressure and temperature. It is possible to measure the total volume of the mixture as a function of time. The initial reactant mixture contents are 10% inert. The following data were obtained at 1 atm and 227 °C.

Time (min)	0	3	6	9	12	15
Volume (cm³)	2.00	2.45	3.00	3.67	4.50	5.51

Determine:
A) The reaction order and the reaction rate coefficient
B) The volume for each time if the feed is composed by pure A.

17) The following first-order gas phase reaction:

$$A \rightarrow 2R$$

was conducted in a closed vessel adapted with a gauge. When the reaction was fixed at 280 °C, the gauge readings were:

Time (sec)	751	∞
Pressure (mmHg)	20.33	30.06

At 305 °C, the gauge readings were:

Time (sec)	320	∞
Pressure (mmHg)	21.29	26.66

Calculate:
A) The reaction rate coefficient at 280 °C
B) The reaction rate coefficient at 305 °C
C) The activation energy and the pre-exponential factor.

18) The following gas phase reaction:

$$A \rightarrow R$$

was studied at 373 and 600 K, and the reaction rate equation was:

$$-r_A = \left[6.093 \times 10^6 e^{-8570/T} (mol/lt)^{0.5} \min^{-1} \right] C_A^{0.5} \quad T \text{ in K}$$

Calculate:

A) The half-life time when the reaction is conducted at $127\,°C$, starting with pure A at 1 atm

B) The time required to reach 90% of conversion if the reaction is performed at 500 K, starting with a blend of 80% of A and 20% of inert at a total pressure of 5 atm

C) The reaction rate at 30 min, starting with pure reactant at 10 atm and a constant temperature of 400 K. The required units for the reaction rate are atmospheres and seconds.

D) The temperature for a half-life time of 25 min starting from $C_{A0} = 1$ mol/lt.

19) The reaction $A \rightarrow 2B$ is carried out at $100\,°C$ at constant pressure of 1 atm in an experimental batch reactor. The following data were obtained starting from pure A.

Time (min)	0	1	2	3	4	5	6	7
V/V_0	1.00	1.20	1.35	1.48	1.58	1.66	1.72	1.78

Time (min)	8	9	10	11	12	13	14
V/V_0	1.82	1.86	1.88	1.91	1.92	1.94	1.95

Determine the reaction order and the reaction rate coefficient.

20) The following decomposition reaction in gas phase is conducted at constant density. Runs 1–5 were done at $100\,°C$, and run 6 at $110\,°C$.

$$A \rightarrow B + 2C$$

Run	1	2	3	4	5	6
Half-life time (min)	4.10	7.70	9.80	1.96	1.30	2.00
C_{A0} (mol/lt)	0.0250	0.0133	0.0100	0.0500	0.0750	0.0250

Determine:

A) The reaction order and the reaction rate coefficient

B) The activation energy.

21) The gas phase dimerization of chlorotrifluoroethylene can be represented by the following reaction:

$$2C_2F_3Cl \rightarrow C_4F_6Cl_2$$

The following data were obtained at $440\,°C$ in a batch reactor at constant density starting from pure reactant:

Time (sec)	0	100	200	300	400	500
Total pressure (kPa)	82.7	71.1	64.4	60.4	56.7	54.8

Determine:

A) The reaction order and the reaction rate coefficient

B) The required time to reach 80% of conversion starting from an initial total pressure of 100 kPa, with a feed composition of 25 mol% of reactant and 75 mo% of N_2.

22) The following data correspond to the decomposition reactor of nitrogen monoxide at 1030 K.

Initial pressure N_2O (mmHg)	82.5	139.0	296.0	360.0
Half-life time (sec)	860	470	255	212

Find the reaction order and the reaction rate coefficient.

23) The following experimental data correspond to the decomposition of ammonia according to:

$$2NH_3 \rightarrow N_2 + 3H_2$$

Time (sec)	100	200	400	600	800	1000
Increase in total pressure (mmHg)	11.0	22.1	44.0	66.3	87.9	110.0

Determine:

A) The reaction order and the reaction rate coefficient, if the reaction starts at 200 mmHg and 600 °C

B) The temperature to obtain 80% of conversion in 500 sec, if the initial concentration is the same as (A) and the activation energy is 40,000 cal/mol.

24) The rate of decomposition of nitrous oxide heated at 990 °C was measured by following the changes in the total pressure with time:

Time (min)	0	30	53	100
Total pressure (mmHg)	200	216	225	236

If the reaction stoichiometry is:

$$2N_2O \rightarrow 2N_2 + O_2$$

Determine:

A) The total pressure to achieve 100% conversion of N_2O

B) The half-time life

C) The reaction order if the half-time life is 52 min, but starting with an initial pressure of 400 mmHg and 990 °C

D) The time to obtain 85% conversion of N_2O.

Irreversible Reactions with Two or Three Components

1) The following reaction between n-propyl bromide and sodium thiosulfate:

$$C_3H_7Br + Na_2S_2O_3 \rightarrow C_3H_7S_2O_3Na + NaBr$$

was followed by titration of the remaining thiosulfate with a solution of 0.02572 N of iodine at 37.5 °C. The following data were obtained by taking 10.02 ml of sample:

Time (sec)	0	1110	2010	3192	5052	7380	11232	78840
I_2 (ml)	37.63	35.20	33.63	31.90	29.86	28.04	26.01	22.24

Determine the reaction rate equation.

2) The following reaction is carried out in gas phase in a variable-density reactor at 585 mmHg and 300 °C. At this temperature, the reaction rate coefficient is 0.5 $(mol/lt)^{-1}sec^{-1}$.

$$A + B \rightarrow 2R + S$$

A) Calculate the required time to obtain 80% conversion of the limiting reactant if the feed molar composition is: 40% of A, 10% of B and 10% of N_2.

B) Under the same conditions as (A), determine the required time to obtain 80% conversion, if the reaction is carried out at constant density.

3) The hydrogenation of benzene was done at 220 °C and 5 atm in a batch reactor at variable density, according to:

$$C_6H_6 + 3H_2 \rightarrow C_6H_{12}$$

For a feed with a molar ratio of $H_2/C_6H_6 = 5$, the following data were obtained:

Time (min)	0.81	1.74	2.78	5.45	9.60	18.53
Conversion (x_A)	0.1	0.2	0.3	0.5	0.7	0.9

A) Calculate the reaction order and the reaction rate coefficient at 220 °C.

B) If the initial reacting volume is 10 lt, determine the volume at 5 min if the feed molar ratio of H_2/C_6H_6 is 3.

C) If the reaction is carried out at constant density and temperature, calculate the conversion of the limiting reactant at 5 min under the conditions of (B).

D) Calculate the reaction rate of limiting reactant at 10 min at constant density and variable density under the conditions of (B).

4) A solution of A is mixed with equal volume of a solution of B, both containing the same number of moles. Then, the following reaction proceeds:

$$A + B \rightarrow R$$

At 1 h, 75% of A is converted. Determine the unconverted amount of A at 2 h, if the reaction follows the kinetics:

A) First order with respect to A and zero order with respect to B

B) First order for both reactants

C) First order with respect to A and second order with respect to B.

5) The following reaction reported the following experimental data:

$$A + B \rightarrow C + D$$

C_A (mol/lt) $\times 10^3$	10.00	7.40	6.34	5.50	4.64
Time (min)	0	3	5	7	10

if the reaction is carried out at 25 °C and $C_{A0} = C_{B0}$. Determine:

A) The reaction rate equation

B) The half-life time at 40 °C if the activation energy is 12.3 kcal/mol.

C) The time to reach 95% conversion at 25 °C, $C_{B0} = 2\, C_{B0}$.

6) The following liquid phase reaction:

$$A + 2B \rightarrow C + 2D$$

is carried out in a batch reactor at 36 °C. The initial concentrations of A and B are: 0.1 and 0.4 mol/lt, respectively. The reaction was followed by analysing the concentration of the product D, which obtained the following data:

Time (min)	0.00	0.60	2.16	3.19	4.50	8.59	18.90
C_D (mol/lt)	0.00	0.02	0.06	0.08	0.10	0.14	0.18

Determine:

A) The reaction order and the reaction rate coefficient

B) The required time to obtain 80% conversion of the limiting reactant, if $C_{A0} = 0.1$ and $C_{B0} = 0.5$ mol/lt.

7) The following gas phase reaction:

$$2A + B \rightarrow R + S$$

was carried out in an isothermal hermetic batch reactor at 727 °C and an initial pressure of 5 atm. If the feed molar composition is 45% A, 41% B and the rest inert, and at 727 °C the value of the reaction rate coefficient is 0.1 (lt/mol) sec, calculate:
A) The required time to reach 75% conversion of the limiting reactant
B) The total pressure at 15 min.

8) The saponification of ethyl acetate in aqueous solution:

$$CH_3COOC_2H_5 + NaOH \rightarrow CH_3COONa + C_2H_5OH$$

was studied by determining the initial reaction rates, and the following data were obtained:

Initial concentration of AcEt (mol/lt)	Initial concentration of NaOH (mol/lt)	Initial reaction rate $\times 10^3$ (mol/lt sec)
0.10	0.10	0.050
0.05	0.10	0.524
0.05	0.05	0.261
0.10	0.05	0.526

Find the individual reaction orders and the reaction rate coefficient.

9) Four runs were done to study the following reaction:

$$A + B \rightarrow S$$

Each run was conducted in a hermetic reactor, and the following initial reaction rates were obtained:

Temperature (°C)	Concentration of A (mol/lt)	Concentration of B (mol/lt)	$(-r_A)$ (mol/lt min)
25	0.5	1.0	0.1886
25	1.0	1.5	0.5659
40	1.5	2.0	2.5159
40	1.0	1.0	0.8386

Determine:

A) The reaction rate equation
B) The value of the reaction rate coefficient at $30\,^\circ C$
C) The required time to reach 90% conversion of the limiting reactant for $C_{A0} = 0.2$ mol/lt and $C_{B0} = 0.5$ mol/lt.

10) In the following reaction:

$$ClO^- + Br^- \rightarrow BrO^- + Cl^-$$

100 ml of $NaClO$ 0.1 N, 48 ml of $NaOH$ 0.5 N and 21 ml of distilled water were mixed. The mixture was immersed in a thermostatic bath at $25\,^\circ C$. 81 ml of a dissolution of 1% KBr was added to the mixture at $25\,^\circ C$. The concentrations of $NaClO$ and KBr in the reacting mixture at $t = 0$ were 0.003230 M and 0.002508 M, respectively, at a pH of 11.28. The following data were obtained:

Time (min)	0.00	3.65	7.65	15.05	26.00	47.60	90.60
Concentration of $BrO^- \times 10^2$ (mol/lt)	0.0	0.0560	0.0953	0.1420	0.1800	0.2117	0.2367

Find the reaction order and the reaction rate coefficient.

11) The decomposition of oxalic acid was studied by its dissolution at 1/40 M in H_2SO_4 at 99.5% at $50\,^\circ C$. Portions of the resulting mixture were separated at different times, and so the required volume of a dissolution of potassium permanganate to react with a portion of 10 ml was measured. The obtained experimental data are:

Time (min)	0	120	240	420	600	900	1440
V (ml)	11.45	9.63	8.11	6.22	4.79	2.97	1.44

Determine the reaction order with respect to the oxalic acid and the reaction rate coefficient.

12) For the liquid phase reaction:

$$A + 2B \rightarrow R + S$$

The following reaction rate equations have been proposed:

$$-r_A = kC_A C_B^2$$

For $C_{Bo}/C_{Ao} = M_{BA} = 2$, the following experimental data were obtained at $70\,^\circ C$ and $C_{A0} = 1$ mol/lt. Find the global reaction order and the reaction rate coefficient.

Time (sec)	10.4	15.8	36.9	86.0	271.2	615.6
x_A	0.17	0.23	0.38	0.54	0.72	0.81

13) The reaction of nitrobenzene was studied at 25 °C:

$$(C_2H_5)_3N + CH_3I \rightarrow [(C_2H_5)_3N^+(CH_3)]I^-$$

and the following experimental data were obtained:

Time (sec)	1200	1800	2400	3600	4500	5400
$C_{(C2H5)3N}$ (mol/lt $\times 10^{-3}$)	8.76	10.66	12.08	13.92	14.76	15.38

The initial concentrations of $(C_2H_5)_3N$ and CH_3I are both 0.0198 mol/lt. If the reaction follows second-order kinetics, calculate the value of the reaction rate coefficient.

14) The following gas phase reaction:

$$A + B \rightarrow R$$

was carried out in a hermetic reactor adapted with a pressure gauge. The experiment started with an equimolar mixture of A and B, and the following data were obtained at 100 °C.

Time (min)	0	5	25	50	100	200	500
P (atm)	1.000	0.963	0.855	0.780	0.654	0.620	0.556

Calculate the required time to obtain 75% of conversion if the reaction is conducted at a constant total pressure of 2 atm, 100 °C and a feed molar composition of 30% A, 60% B and 10% inert.

15) The kinetics of the reaction of methanol with 3,3-dicarbazil phenyl-methyl ion (green colorant) in a water–ketone dissolution was investigated spectrophotometrically at 25 °C. It was found that for an initial solution of 0.25 M in methanol, the difference between the absorbance of the dissolution at any time (A) and the absorbance of the dissolution when the reaction has reached the equilibrium (A_∞) at 730 nm varies with respect to time as follows. When the concentration of NaOH is kept at 2.03×10^{-2} M:

Time (min)	2.7	8.7	14.7	21.7	33.7
$A-A_\infty$	0.562	0.243	0.111	0.045	0.010

The absorbance is directly proportional to the concentration of the absorbent species. If in the light that passes through the simple,

P is the transmitted energy and P_0 is the incident energy, the absorbance is given by:

$$A = \log \frac{P_o}{P}$$

Demonstrate that the reaction follows first-order kinetics with respect to the ion, and calculate the reaction rate coefficient.

16) The gas phase reaction between A and B is carried out by isothermal measures of the half-life time at different initial partial pressures of reactants:

P_{Ao} (mmHg)	500	125	250	250
P_{Bo} (mmHg)	10	15	10	20
$(t_{1/2})_B$ (min)	80	213	160	80

Find the reaction rate expression.

17) The following gas phase reaction:

$$2A + B \rightarrow R + S$$

was conducted in a hermetic reactor operated at a constant temperature of $727\,^\circ C$ and initial pressure of 5 atm. If the feed molar composition is 41% A, 41% B and the rest inert, and the value of the reaction rate coefficient at $727\,^\circ C$ is 0.1 lt/mol sec, calculate:

A) The required time to obtain 75% of conversion of the limiting reactant

B) The total pressure at 15 min.

18) The following reaction:

was studied by preparing an initial dissolution of 0.12 M in ethylene oxide and 0.007574 M in perchloric acid. The reaction extent was followed dilatometrically (i.e. by measuring the volume of the dissolution as a function of time). The following data were obtained at $20\,^\circ C$:

Time (min)	0	30	60	135	300	∞
Reading of the dilatometer	18.48	18.05	17.62	16.71	15.22	12.29

Determine the reaction order with respect to ethylene oxide, and calculate the reaction rate coefficient.

19) For the following reaction, the next dependency of the reaction rate coefficient (k) with ionic strength (I) was found:

$$Co(NH_3)_5NO_2^{+2} + OH^- \rightarrow Co(NH_3)_5OH^{+2} + NO_2^-$$

I	2.34	5.61	8.10	11.22	11.73	16.90
$5 + \log k$	1.7640	1.7130	1.6800	1.6467	1.6418	1.5990

Calculate the reaction rate coefficient k_o corresponding to the zero ionic strength.

20) The reaction in liquid phase:

$$A + B \rightarrow R + S$$

is carried out at a constant temperature of 25 °C in a batch reactor with $C_{Ao} = 0.054$ mol/lt and $C_{Bo} = 0.106$ mol/lt. The following experimental data were obtained:

Time (min)	174	418	426	444	1150	1440	1510	1660
x_A	0.203	0.335	0.350	0.383	0.588	0.618	0.638	0.655

Find the reaction order and the reaction rate coefficient.

21) The reaction in liquid phase is carried out at 100 °C in a batch reactor:

$$A + B \rightarrow R$$

The following experimental data were obtained when the reacting mixture contains 3 lt of a solution with 2 moles of A and 2 lt of another solution with 1 mol of B.

Time (min)	10	20	30	40	50
C_A (mol/lt)	0.375	0.354	0.336	0.322	0.309

Find the reaction rate expression.

22) The following reaction was studied in liquid phase at 25 °C:

$$CH_3OH + (C_6H_5)_3CCl \rightarrow (C_6H_5)_3COCH_3 + HCl$$

The reaction order with respect to methanol is two, and it is one for triphenylmethyl chloride, with a value of the reaction rate coefficient of 0.27 $(mol/lt)^{-2}min^{-1}$. The initial concentrations were 0.054 and 0.106 mol/lt, respectively. Calculate the required time to obtain 50% conversion of the limiting reactant.

23) The following liquid phase reaction:

$$A + B \rightarrow R + S$$

was conducted in a hermetic reactor at a constant temperature of 50 °C. The following experimental data were obtained for $C_{Bo}/C_{Ao} = 3$:

Time (h)	0	1	2	4	6
C_A (mol/lt)	500	225	110	30	10

Determine:
A) The reaction rate expression
B) The required time to reduce the initial concentration to a half of its value at 50 °C.

24) The following reaction rate expression was found experimentally at 80 °C in a batch reactor:

$$-r_A = 0.005 C_A C_B^2$$

$$A + B \rightarrow R + S$$

$k = 0.005 \ (\text{mol/lt})^{-2}\text{min}^{-1}$

If $C_{Ao} = C_{Bo} = 1$ mol/lt, calculate the conversion at 80 °C and 5 h.

25) The following reaction was studied in a batch reactor at a constant temperature of 80 °C and $C_{Ao} = 2$ mol/lt and $C_{Bo} = 5$ mol/lt:

$$A + B \rightarrow Products$$

Time (h)	3.10	5.98	8.29	20.74	27.19	32.92	41.25
C_A (mol/lt)	1.60	1.32	1.14	0.56	0.40	0.30	0.20

A) Find the reaction rate expression.
B) If the activation energy is 20,000 cal/mol, find the reaction rate expression at 130 °C.
C) If the reaction is conducted at 80 °C with an excess of B ($M_{BA} = 5$), plot the conversion (x_A) against time for the first 6 h of the reaction.
D) Calculate the half-life time at 80 °C.

26) The gas phase reaction:

$$A + B \rightarrow R + 2S$$

was carried out isothermally at 100 °C in a hermetic reactor. The reaction rate expression is:

$$-r_A = 0.064 C_A C_B$$

where $k[=](\text{mol/lt})^{-1}\text{min}^{-1}$.

Calculate the total pressure of the reactor at 15 min, if $P_{Ao} = P_{Bo} = 2$ atm, and $P_I = 1$ atm (inert).

27) The following elemental reaction:

$$A + B \rightarrow C$$

is carried out at industrial scale. Normally, the reactants are pre-heated to the reaction temperature and then are loaded to the reactor in equimolar composition. The conversion obtained at 10 h is 99%. To increase the productivity, it is assumed that the feed composition should be changed to 2 moles of B for each mol of A, and hence the reaction may be faster, without changing the final conversion. The reacting volume in both cases would be the same and constant. Is this proposed approach correct?

28) Four runs were done to study the following reaction:

$$A + B \rightarrow S$$

Each run was carried out in a hermetic reactor, and the following experimental data were obtained:

Experiment	Temperature (°C)	C_{Ao}	C_{Bo}	$-r_A$ (mol/lt min)
1	25	0.5	1.0	0.1886
2	25	1.0	1.5	0.5659
3	40	1.5	2.0	2.5159
4	40	1.0	1.0	0.8386

Calculate:
A) The reaction rate expression
B) The reaction rate coefficient at 30 °C
C) The required time to obtain 90% of conversion of the limiting reactant for initial concentrations of 2 mol/lt of A and 0.5 mol/lt of B at 30 °C.

29) The following reaction:

$$2NO + 2H_2 \rightarrow N_2 + 2H_2O$$

was studied in gas phase at 826 °C by measuring initial reaction rates at different initial partial pressures. Find the reaction order with respect to each reactant.

Experiment	$P_{o\ H2}$ (mmHg)	$P_{o\ NO}$ (mmHg)	$-r_{Ao}$ (mmHg/sec)
1	400	300	1.03
2	400	152	0.25
3	289	400	1.60
4	147	400	0.79

Reversible Reactions

1) The following reaction was carried out in liquid phase at 14 °C:

$$A \underset{k_2}{\overset{k_1}{\rightleftharpoons}} R$$

$$(-r_A) = k_1 C_A - k_2 C_R$$

From the following data, calculate $(k_1 + k_2)$, if $C_{Ro} = 0$. Note that optical rotation is an additive property of each of the reacting species.

Time (sec)	0.0	10.8	18.0	25.2	86.4	259.2	∞
Reading of the polarimeter (degrees)	189.0	169.0	156.0	146.0	84.5	37.3	31.3

2) The following data of the aqueous solution reaction at a constant temperature were obtained:

$$A + B \underset{k_2}{\overset{k_1}{\rightleftharpoons}} R$$

Time (min)	0	1	2	3	5	7	10	20	∞
C_R (mol/lt)	0.00000	0.00328	0.00598	0.00820	0.01152	0.01376	0.01582	0.01804	0.01840

If $C_{Ao} = 0.02$ mol/lt, $C_{Bo} = 10$ mol/lt and $C_{Ro} = 0$, find the reaction rate expression.

3) The following reaction:

$$A + B \underset{k_2}{\overset{k_1}{\rightleftharpoons}} 2R$$

was studied at $60\,°C$ by mixing $100\,ml$ of a solution of A with a concentration of $1.6\,mol/lt$, and the following data were obtained:

Time (min)	0	10	30	50	80	100	∞
C_R (mol/lt)	0.0000	0.2640	0.5920	0.7632	0.8960	0.9408	1.0240

Determine:

A) The reaction rate expression

B) The required time to obtain 30% of conversion of the limiting reactant if $C_{Ao} = 0.8\,mol/lt$, $C_{Bo} = 1.4\,mol/lt$ and $C_{Ro} = 0.1\,mol/lt$.

4) For the liquid phase reaction:

$$A \underset{k_2}{\overset{k_1}{\rightleftharpoons}} 2R$$

the following data were obtained at $28\,°C$, starting with pure A with $C_{AO} = 0.05\,mol/lt$.

Time (min)	5	10	15	20	25	∞
C_R (mol/lt)	0.028	0.046	0.058	0.066	0.072	0.082

A) Find the reaction rate expression.

B) If $C_{AO} = C_{RO}$, calculate the time to obtain $x_A = 0.8$.

5) The reversible isomerization reaction in gas phase of A is carried out in a batch reactor at constant temperature. With initial concentrations of $1.6\,mol/lt$ of A and $0.2\,mol/lt$ of the product R, the following experimental data were obtained:

Time (min)	10	20	30	40	50	60	∞
C_R (mol/lt)	0.643	0.931	1.126	1.243	1.328	1.379	1.480

A) Find the reaction rate expression.

B) Calculate the equilibrium constant.

C) Calculate the concentration of R at 2 min.

6) A solution containing the substance A in a concentration of $1.6\,mol/lt$ and the product R in a concentration of $0.3\,mol/lt$ is rapidly heated to the reaction temperature, which is maintained constant. The following concentrations of A are measured at different times:

Time (min)	0	0.5	1.0	1.5	2.0	3.0	4.0	5.0	10.0
C_A (mol/lt)	0.800	0.670	0.600	0.563	0.543	0.527	0.522	0.520	0.500

Find the reversible reaction rate expression, assuming first order of the forward and reverse reactions.

7) The following data were obtained of a reaction in aqueous solution at 20 °C. Starting with a solution with a concentration of A of 0.1 mol/lt, determine the reaction rate expression.

$$A \underset{k_2}{\overset{k_1}{\rightleftarrows}} R$$

Time (min)	19.31	43.61	119.98	189.27	∞
x_A	0.1	0.2	0.4	0.5	0.675

8) The production of propionic acid is carried by acidifying an aqueous solution of sodium salt according to the following reaction:

$$C_2H_5COONa + HCl \underset{k_2}{\overset{k_1}{\rightleftarrows}} C_2H_5COOH + NaCl$$

The reaction can be represented by a second-order kinetics in both path reactions. The following data were obtained by taking a sample of 10 ml of reacting solution at different times and neutralizing the unreacted HCl with a solution of $NaOH$ 0.515 N. The initial concentrations of HCl and C_2H_5COONa are the same.

Time (min)	0	12	20	30	50	∞
NaOH solution (ml)	52.5	32.1	23.5	18.9	14.4	10.5

Determine:
A) The equilibrium constant
B) The values of the reaction rate coefficients
C) The required time to obtain 73.75% conversion if the initial concentration is 20% higher.

9) The isomerization of 1,2 dimethylcyclopropane (from cis- to trans-) at 435 °C reported the following data:

Time (sec)	0	45	90	255	270	360	495
Concentration of cis- (mol/lt)	100.0	89.2	91.1	62.3	58.2	50.7	43.5

After a long time, it was found that the concentration of the reactant is 30 mol/lt. Determine the reaction rate expression.

10) The isomerization in gas phase:

$$A \underset{k_2}{\overset{k_1}{\rightleftarrows}} R$$

was studied in a wide range of temperatures (100–400 °C), and the following Arrhenius expressions were found for k_1 and k_2.

$$k_1 = 6 \times 10^{14} e^{-40000/RT} \, [=] \min^{-1}$$

$$k_2 = 3.7 \times 10^{12} e^{-37000/RT} \, [=] \min^{-1}$$

Calculate:
A) The equilibrium conversion at 200 °C
B) The required time to obtain 80% of the equilibrium conversion at 300 °C.

11) For the reaction:

$$A + B \underset{k_2}{\overset{k_1}{\rightleftarrows}} C$$

the following data were obtained at 22.9 °C with $C_{Ao} = C_{Bo} = 5.5$ gmol/lt. Find k_1 and k_2.

Time (sec)	0	1680	2880	7620	10800	19080	∞
C_C (gmol/lt)	0	0.69	1.38	3.31	4.11	5.15	5.80

12) The following data correspond to a first-order reaction in the forward route and second-order in the reverse route. The initial concentration of each reactant is the same, 1.2 mol/lt, and there is not a product at the beginning of the reaction. Calculate:
A) The reaction rate coefficients
B) The equilibrium constant
C) The required time to obtain 60% conversion.

Moles of A	0.00	0.05	0.10	0.15	0.20	0.30	0.40	0.50	0.60	0.70	0.80
$-r_A$ (mol/lt min)		2.16	1.96	1.79	1.60	1.44	1.12	0.84	0.59	0.38	0.17	0.00

13) The following reaction:

$$A \underset{k_2}{\overset{k_1}{\rightleftarrows}} R$$

was carried out in a batch reactor with $C_{Ao} = 1.6$ mol/lt and $C_{Bo} = 1$ mol/lt. Prior to the reaction, the solutions are mixed with the initial concentration indicated in the table in the same proportion. The experimental data were obtained at 50 °C. Calculate the equilibrium constant and the values of k_1 and k_2.

Time (min)	0	0.5	1.0	1.5	2.0	3.0	4.0	∞
C_A (mol/lt)	0.80	0.67	0.60	0.563	0.543	0.527	0.522	0.520

14) The following data correspond to the following reaction conducted at 25 °C:

$$A \underset{k_2}{\overset{k_1}{\rightleftharpoons}} R + S$$

where R is in great excess and $C_{Ao} = 182.5 \text{ mol/m}^3$.
A) Find the reaction rate expression.
B) Calculate the required time to obtain 95% of the equilibrium conversion.

Time (sec)	0	126	300	480	720	1320	∞
C_S (mol/m³)	0	24.10	49.91	70.80	90.00	115.50	132.80

15) The following gas phase reaction was carried out in a hermetic vessel:

$$A \underset{k_2}{\overset{k_1}{\rightleftharpoons}} R + S$$

The initial partial pressures were: $p_{Ao} = 0.2$ atm, $p_{Ro} = 24.8$ atm and $p_{So} = 0$. The reactor was kept in an oil bath at 157 °C. The following experimental data were obtained. Find the reaction rate expression.

Time (min)	2	4	8	16	20	∞
P_S (atm)	0.03142	0.05504	0.08615	0.11366	0.11927	0.12656

16) The following reaction:

$$2A \underset{k_2}{\overset{k_1}{\rightleftharpoons}} R + S$$

is conducted in gas phase, with a constant pressure of 1 atm and 1033 K. The reaction rate expression at such conditions is:

$$-r_A = 14.96 \times 10^6 e^{-15200/_R T} \left(p_A^2 - \frac{p_R p_S}{K_p} \right)$$

Where T is the temperature in K; p_A, p_R and p_S are the partial pressures in atm; K_p is the equilibrium constant as a function of partial pressures; and $(-r_A)$ is the reaction rate in mol/lt h, the value of which is 0.312 at 1033 K.

Calculate:

A) The values of k_1 and k_2
B) The equilibrium conversion at 1033 K
C) The required time to obtain 90% of the equilibrium conversion.

17) The following second-order reversible liquid phase reaction is carried out in a batch reaction with $C_{Ao} = 0.5$ mol/lt and $C_{Ro} = 0$:

$$A \underset{k_2}{\overset{k_1}{\rightleftharpoons}} 2R$$

Find the reaction rate expression if the equilibrium conversion is 66.7% and the half-life time is 8 min.

18) For the liquid phase reaction:

$$A \underset{k_2}{\overset{k_1}{\rightleftharpoons}} 2R$$

the following experimental data were obtained at 25 °C with pure A and $C_{Ao} = 0.05$ gmol/lt.

Time (min)	5	10	15	20	25	∞
x_A	0.28	0.46	0.58	0.66	0.72	0.82

A) Find the reaction rate expression.
B) If $C_{AO} = C_{Ro}$, calculate the required time to obtain a conversion of 50%.

19) The conversion of A into R is represented by the reaction:

$$A \underset{k_2}{\overset{k_1}{\rightleftharpoons}} R$$

The reaction was studied by using a solution with 18.23 gmol/lt of A, obtaining the following data:

Time (min)	0	21	36	50	65	80	100	∞
C_R (gmol/lt)	0	2.41	3.73	4.96	6.10	7.08	8.11	13.28

Determine:

A) The equilibrium conversion and the equilibrium constant
B) The reaction rate coefficients
C) The conversion at 70 min
D) If $C_{Ao} = C_{Ro}$, the required time to obtain a conversion of 50%.

20) Butadiene is obtained by dehydrogenation of 1-butene in the presence of steam:

$$C_4H_8 \underset{k_2}{\overset{k_1}{\rightleftharpoons}} C_4H_6 + H_2$$

$K_p = 0.252$ atm at 900 °C

$$k_1 = 1.5 \times 10^8 e^{-33600/_R T} \, [=] \, \text{min}^{-1}$$

If the feed composition contains a mixture of steam and 1-butene with a molar ratio of 15/1 at 900 K and 1 atm, calculate:
A) The equilibrium conversion
B) The required time to obtain 90% of the equilibrium conversion.

21) The following reaction was studied in aqueous solution:

$$CH_3COCH_3 \;+\; HCN \;\rightleftharpoons\; (CH_3)_2C\overset{\displaystyle CN}{\underset{\displaystyle OH}{\Big\langle}}$$

With initial concentrations of 0.0758 N of HCN and 0.1164 N of ketone, the following data were obtained:

Time (min)	4.37	73.2	172.5	265.4	346.7	434.4
C_{HCN} (N)	0.0748	0.0710	0.0655	0.0610	0.0584	0.0557

Find the reaction rate expression if $K_c = 13.87$ lt/mol.

22) The gas phase reaction:

$$A + B \underset{k_2}{\overset{k_1}{\rightleftharpoons}} R$$

is represented by the following reaction rate expression:

$$r_A = k_1 C_A C_B - k_2 C_R$$

If the initial total pressure is 1 atm, the feed is equimolar ($y_{Ao} = y_{Bo}$) and $C_{Ro} = 0$, calculate the required time to obtain a conversion of 90% of the equilibrium conversion at 200 °C. The reaction rate coefficients are given by:

$$k_1 = 1.5 \times 10^9 e^{-15200/_{RT}}$$

$$k_2 = 2.2 \times 10^{12} e^{-33600/_{RT}}$$

where T is the temperature in K and $R = 1.9872$ cal/mol K.

23) The following reaction is carried out in aqueous solution with initial concentrations of A of 0.180 gmol/lt, P of 54 gmol/lt and S of 0.

$$A \underset{k_2}{\overset{k_1}{\rightleftharpoons}} S + P$$

The experimental data are:

Time (min)	0	35	45	65	150	250	350	450	∞
x_A	0	0.21	0.26	0.35	0.60	0.74	0.80	0.83	0.85

A) Find the reaction rate expression.
B) Calculate the reaction rate coefficients.

24) The following reaction was studied at 20 °C and 1 atm.

$$A \underset{k_2}{\overset{k_1}{\rightleftharpoons}} R$$

Starting with pure A, the following data were obtained:

Time (min)	10	20	30	40	50	60	70	80	90	100	120	130
$C_A \times 10^2$ (gmol/lt)	3.45	2.98	2.65	2.44	2.29	2.18	2.11	2.06	2.03	2.00	1.96	1.96

The value of the equilibrium constant at 20 °C is 0.0995. Determine:
A) The reaction rate expression
B) The required time to obtain a conversion of 10%
C) The conversion at 2.5 h.

Complex Reactions

1) For the following reaction:

$$A \xrightarrow{k_1} R \xrightarrow{k_2} S$$

with $k_1/k_2 = 3$ and $k_1 = 2\,\mathrm{h^{-1}}$, calculate:
A) The required time to obtain the maximum concentration of R
B) The conversion of A at the time in which the maximum concentration of R is achieved
C) The yields of R and S at the time calculated in (B).

2) For the following reaction:

$$A \xrightarrow{k_1} R$$
$$A \xrightarrow{k_2} S$$

For $k_1 = 2\,h^{-1}$ and $k_2 = 0.667\,h^{-1}$, calculate:

A) The yield of R and S, when conversion of A is 98%

B) The required time to obtain 95% of conversion of A.

3) The following experimental data were obtained at $12\,°C$ for the reaction:

$$A \xrightarrow{\;k_1\;} R \xrightarrow{\;k_2\;} S$$

$E_{A1} = 18{,}000\ \text{cal/mol}$
$E_{A2} = 16{,}500\ \text{cal/mol}$
$k_2 = 0.0135\ \text{min}^{-1}$

Time (min)	0	4	12	28	60
C_A (mol/lt)	0.8000	0.5948	0.3288	0.1256	0.0094
C_R (mol/lt)	0.0000	0.1995	0.4299	0.5475	0.4237

Calculate:

A) The value of k_1 at $12\,°C$

B) The concentration of S for each time at $12\,°C$

C) The concentration of S for each time at $20\,°C$.

4) The reactant A decomposes according to:

$$A \xrightarrow{\;k_1\;} R$$

$$A \xrightarrow{\;k_2\;} S$$

Calculate:

A) k_1 and k_2 if, in 15 min, a yield of R of 60% and a concentration of A of 0.05 gmol/lt are obtained, starting with concentrations of 0.05 gmol/lt of R and 0.5 gmol/lt of A.

B) If the reaction was in series ($A \xrightarrow{\;k_1\;} R \xrightarrow{\;k_2\;} S$), with the values of k_1 and k_2 calculated in (A), but feeding only A with a concentration of 0.3 gmol/lt, calculate the required time to obtain a yield of S of 80%, the maximum concentration of R and the time to reach such a concentration.

5) For the following hydrogenation reaction:

$$A \xrightarrow{\;k_1\;} R \xrightarrow{\;k_2\;} S$$

the concentration of R has a maximum value at 5 min, starting with an initial concentration of A of 0.2 gmol/lt. If $k_1 = 0.215\ \text{min}^{-1}$, calculate the maximum concentration of R and the time at which it is reached, and the concentrations of A and S at such time.

6) For the following liquid phase reaction:

$$A \xrightarrow{k_1} R$$
$$A \xrightarrow{k_2} S$$

with $C_{A0} = 100$ mol/lt and $C_{R0} = C_{S0} = 0$, the following data were obtained:

C_A (mol/lt)	90	80	70	60	50	40
C_R (mol/lt)	7	13	19	22	25	27

Calculate the value of k_1/k_2.

7) For the following reactions:

$$A \xrightarrow{k_1} R$$
$$A \xrightarrow{k_2} S$$

the following experimental data were obtained:

Time (min)	0	10	20	30	40	50	60
C_A (mol/lt)	61.9	55.4	49.8	45.1	39.7	36.8	32.5

At the end of the reaction, yields of R and S were 75% and 25%, respectively. Determine the value of k_1 and k_2.

8) The following liquid phase reaction:

$$A \xrightarrow{k_1} 2R$$
$$A \xrightarrow{k_2} S$$

was studied at $100\,^{\circ}C$ in a batch reactor, and the yields of R and S were 50 and 20%, respectively, at 10 min for an initial concentration of A of 0.14 mol/lt.

Determine:
A) The values of k_1 and k_2
B) The yields of R and S at 5 min for $C_{A0} = 0.2$ mol/lt.

9) For the following reaction:

$$A \xrightarrow{k_1} R \xrightarrow{k_2} S$$
$$k_1 = 0.35 \text{ h}^{-1}$$
$$k_2 = 0.13 \text{ h}^{-1}$$

calculate the following for an initial concentration of A of 4 mol/lt:
A) The required time for the maximum concentration of R
B) The concentrations of A and S at such time

C) The maximum concentration of R

D) The time at which the concentrations of A and S are the same.

10) The synthesis of monochlorobenzene (MCB) is carried out as follows:

$$C_6H_6 + Cl_2 \rightarrow MCB + HCl$$

$$MCB + Cl_2 \rightarrow DCB + HCl$$

This reaction system behaves as an in-series reaction:

$$A \xrightarrow{k_1} R \xrightarrow{k_2} S$$

For an initial concentration of benzene of 6 mol/lt, and $k_2/k_1 = 5$ and $k_1 = 0.75\,h^{-1}$, calculate:

A) The required time to obtain the maximum concentration of MCB

B) The concentrations of benzene and dichlorobenzene (DCB) corresponding to the maximum concentration of MCB

C) The maximum concentration of MCB

D) The concentration of DCB at 2 h.

11) The following reactions were carried out in an aqueous solution with equal initial concentrations of A and B:

$$A + B \xrightarrow{k_1} R$$

$$A + B \xrightarrow{k_2} S$$

$$A + B \xrightarrow{k_3} T$$

The yields of R, S and T were 53%, 31% and 17%, respectively, at the end of the reaction. The following experimental data were obtained:

C_A (mol/lt)	2.000	1.394	1.069	0.816	0.513
Time (min)	0	1.5	3.0	5.0	10.0

Calculate the values of k_1, k_2 and k_3.

12) 500 ml of reactant A and 500 ml of reactant B were loaded to a batch reactor. It is believed that these compounds react according to the following scheme:

$$A + B \xrightarrow{k_1} R + S$$

$$A + B \xrightarrow{k_2} T$$

A) Find the values of k_1 and k_2, if the yield of A is 30% in the first reaction and 70% in the second reaction.

B) Calculate the time to obtain 50% of yield of A.

13) In the following elemental reaction:

$$A \xrightarrow{k_1} R \xrightarrow{k_2} S$$

the following data were obtained at $40\,°C$ and $C_{Ao} = 6.439 \times 10^{-3}$ gmol/lt.

t (sec)	390	777	1195	3155
x_A	0.1538	0.2825	0.4002	0.7488

If the required time to obtain the maximum concentration of R is 1480.5 min, calculate:
A) The values of k_1 and k_2
B) The maximum concentration of R (C_R^{max})
C) The concentration of A and S (C_A^* and C_S^*) when C_R^{max} is reached.

14) In the following reaction:

$$A \xrightarrow{k_1} R \xrightarrow{k_2} S$$

the following experimental data were obtained:

t (min)	C_A (gmol/lt)	C_R (gmol/lt)	C_S (gmol/lt)
0	100	0	0
0.2	90.5	8.5	1.0
0.4	81.9	14.6	3.5
0.6	74.1	19.3	6.6
0.8	67.0	23.0	10.0
1.0	60.7	26.0	13.3
1.5	47.2	31.8	21.0
2.0	36.8	35.9	27.3
2.5	28.7	39.7	31.6
3.0	22.3	41.9	35.8
4.0	13.5	44.8	41.7
5.0	8.2	46.9	44.9
6.0	5.0	48.1	46.9
7.0	3.0	48.6	48.4
8.0	1.8	49.3	49.1
10.0	0.7	49.7	49.6

Calculate the values of the reaction rate coefficients k_1 and k_2.

15) The following liquid phase reaction was studied at 51 °C:

$$A + B \xrightarrow{k_1} R$$

$$A + B \xrightarrow{k_2} S$$

Using equimolar feed composition of A and B, the following data were obtained:

t (min)	0	4.5	6.0	9.0	10.5	24.0	27.5	30.0
C_A (gmol/lt)	0.181	0.141	0.131	0.119	0.111	0.0683	0.0644	0.0603

Calculate the values of k_1 and k_2.

Nomenclature

A	Area. Frequency or pre-exponential factor. Integration constant. Chemical species
a	Stoichiometric coefficient of reactant A
a_r	Constant
a_0	Parameter of regression analysis
a_i	Parameter of regression analysis. Polynomial constants
A_i	Chemical formula of species i
B	Chemical species. Integration constant
b	Intercept of the straight line. Stoichiometric coefficient of reactant B
b_r	Constant
C	Integration constant
C_i	Concentration of species i
C_I	Integration constant
C_i^{max}	Concentration at equilibrium of species i
D_1	Discriminant of quadratic equation
E	Specific enzyme of the reaction
E_A	Activation energy
E_{A1}	Activation energy of forward reaction
E_{A2}	Activation energy of reverse reaction
ES	Intermediate complex composed by enzyme and substrate
FCC	Fluid catalytic cracking
F_i	Molar flowrate of species i
f	Parameter of the equation of L'Hôpital's rule
f_i	Fugacity of species i
f'	Parameter of the equation of L'Hôpital's rule
G	Total Gibbs free energy
g	Parameter of the equation of L'Hôpital's rule
g'	Parameter of the equation of L'Hôpital's rule
IQ	Interquartile range
K_m	Constant of Michaelis–Menten

Chemical Reaction Kinetics: Concepts, Methods and Case Studies, First Edition.
Jorge Ancheyta.
© 2017 John Wiley & Sons Ltd. Published 2017 by John Wiley & Sons Ltd.

K	Equilibrum constant
K_c	Equilibrium constant based on concentrations
K_e	Equilibrium constant
K_i	Adsorption constant
K_p	Equilibrium constant based on partial pressures
K_y	Equilibrium constant based on mole fractions
k_c	Reaction rate coefficient as function of concentration
k_d	Deactivation constant
k_{ex}	Constant
k_i	Reaction rate coefficient
k_m	Average reaction rate coefficient
k_p	Reaction rate coefficient as function of partial pressure
k_{v1}	Constant
k_{v2}	Constant
k_1	Reaction rate coefficient of the forward reaction
k_{-1}	Reaction rate coefficient of the reverse reaction
k_2	Reaction rate coefficient of the reverse reaction
k'	Constant
k^*	Constant
$k_{\lambda i}$	Constant
L	Capillary length at time t
$LHSV$	Liquid hourly space velocity
LIF	Lower inner fence
L_o	Capillary length at time zero ($t = 0$)
LOF	Lower outer fence
L_∞	Capillary length that does not change with time
M_{ij}	Feed molar ratio of i with respect to j
MW	Molecular weight
m	Exponent for temperature effect. Slope of the straight line
m_K	Slope of the straight line as function of K
$m_{\chi Ae}$	Slope of the straight line as function of x_{Ae}
n_i	Number of moles of species i. Reaction order of reaction i
n_t	Total number of moles
n	Global reaction order. Number of data. Polynomial order
OF	Objective function
P	Total pressure. Integration parameter. Product of reaction
P_{atm}	Atmospheric pressure
p_i	Partial pressure of species i
Q	Integration parameter
Q_1	Lower quartile
Q_3	Upper quartile
R	Universal gas constant. Chemical species

r	Correlation coefficient. Stoichiometric coefficient of product R. Reaction rate
r_i	Reaction rate of species i
r^2	Determination coefficient or correlation coefficient squared
R_{ij}	Ratio of reaction rate coefficient of species i with respect to j
RM	Reparameterization
ROM	Reduction of orders of magnitude (method)
S	Substrate. Chemical species
S_{ij}	Selectivity of species i with respect to j
S_i	Weight of stream i ($i = 1, 2, 3, 4$)
SIF	Superior inner fence
SOF	Superior outer fence
SSE	Sum of square errors
SV	Space velocity
s	Stoichiometric coefficient of product S
TM	Traditional method
T_m	Average temperature
t	Time
t_r	Constant
$t_{1/2}$	Half-life time
$t_{1/m}$	Time necessary for the initial concentration to reduce to $1/m$ of its value
t^*	Time to reach the maximum concentration
V	Volume of reacting fluid
V_{max}	Maximum reaction rate
v_o	Initial volumetric flowrate
x_A	Conversion of reactant A
x_{Ae}	Equilibrium conversion
x_i	Molar fractional conversion of species i
x_1	Variable for regression analysis
x_2	Variable for regression analysis
$WHSV$	Weight hourly space velocity
w_i	Weighting factor
w_t	Total weight
Y_i	Yield of species i
y	Variable for regression analysis
y_i	Mole fraction of species i
y_i'	Time derivatives
y_{iC}	Mass fraction of carbon on stream i ($i = 1, 3, 4$)
y_{iH2}	Mass fraction of hydrogen on stream i ($i = 1, 3, 4$)
y_{iN2}	Mass fraction of nitrogen on stream i ($i = 1, 3, 4$)
y_{iS}	Mass fraction of sulphur on stream i ($i = 1, 3, 4$)

y_{wi} Weight fraction of species i
y_1 Yield of gas oil
y_2 Yield of gasoline
y_3 Yield of gas plus coke
y_{31} Yield of gas
y_{32} Yield of coke

Greek Letters

α Individual reaction order
β Individual reaction order
γ Individual reaction order
γ_I Fugacity coefficient
ΔH^o Change of enthalpy of the reaction
Δn Change in the number of moles
ε_A Molar expansion factor
ξ Reaction extent
λ_i Physical property of species i
λ Physical property at time t
λ_o Physical property at time zero $(t = 0)$
λ_∞ Physical property that does not change with time
$\xi_i{}'$ Volumetric reaction extent
ξ_i^{max} Maximum reaction extent of species i
ρ_i Density of species i
σ_{Siexp} Standard deviation of experimental data of stream i $(i = 1, 2, 3, 4)$
τ Space–time
υ_i Stoichiometric number of species i
ϕ Catalyst decay function
Ω Electric resistance at time t
Ω_o Electric resistance at zero time $(t = 0)$
Ω_∞ Electric resistance that does not change with time

Subindex

0   Property at zero time
p   Average
calc   Calculated
exp   Experimental
1   Liquid feed stream
2   Input gas stream
3   Liquid product stream
4   Gas product stream

References

Ancheyta, J., Angeles, M.J., Macías, M.J., Marroquín, G., Morales, R. 2002. Changes in apparent reaction order and activation energy in the hydrodesulfurization of real feedstocks. *Energy Fuels.* **16**:189–193.

Ancheyta, J., López, F., Aguilar, E. 1999. 5-Lump kinetic model for gas oil catalytic cracking. *Appl. Cat. A.* **177**:227–235.

Ancheyta, J., López, F., Aguilar, E., Moreno, J.C. 1997. A strategy for kinetic parameter estimation in the fluid catalytic cracking process. *Ind. Eng. Chem. Res.* **36**:5170–5174.

Ancheyta, J., Rodríguez, M.A., Sánchez, S. 2005. Kinetic modeling of hydrocracking of heavy oil fractions: a review. *Catal. Today.* **109**:76–92.

Angeles, M.J., Leyva, C., Ancheyta, J., Ramírez, S. 2014. A review of experimental procedures for heavy oil hydrocracking with dispersed catalyst. *Catal. Today.* **220–222**:274–294.

Arrhenius, S. 1889. Über die Reaktionsgeschwindigkeit bei der Inversion von Rohrzucker durch Säuren. *Z. Phys. Chem.* **4**:226–248.

Avery, H.E. 1983. *Basic reaction kinetics and mechanisms.* MacMillan, London.

Bacaud, R., Pessayre, S., Vrinat, M. 2002. Cinética de reacciones de hidrodesulfuración. XVII Simposio Iberoam. Catal., Isla de Margarita, Venezuela, September.

Bagajewicz, M.J., Cabrera, E. 2003. Data reconciliation in gas pipeline systems. *Ind. Eng. Chem. Res.* **42**:5596–5606.

Basak, K., Abhilash, K.S., Ganguly, S., Saraf, D.N. 2002. On-line optimization of a crude distillation unit with constraints on product properties. *Ind. Eng. Chem. Res.* **41**:1557–1568.

Blanding, F.H. 1953. Reaction rates in catalytic cracking of petroleum. *Ind. Eng. Chem.* **45**:1186–1197.

Buchaly, C., Kreis, P., Gorak, A. 2012. n-Propyl propionate synthesis via catalytic distillation-experimental investigation in pilot-scale. *Ind. Eng. Chem. Res.* **51**:891–899.

Butt, J.B. 1980. *Reaction kinetics and reactor design.* Prentice-Hall, Englewood Cliffs, NJ.

Callejas, M.A., Martínez, M.T. 1999. Hydrocracking of a Maya residue: kinetics and product yield distributions. *Ind. Eng. Chem. Res.* **38**:3285–3289.

Chapra, S.C., Canale, R.P. 1990. *Numerical methods for engineers,* 2nd ed. McGraw-Hill Education, Columbus, OH.

Chen, N.H., Aris, R. 1992. Determination of Arrhenius constants by linear and nonlinear fitting. *AIChE J.* **38**:626–628.

Chopey, N.P. 1994. *Handbook of chemical engineering calculations,* 2nd ed. McGraw-Hill, New York.

Christensen, H.N., Palmer, G.A. 1980. *Enzyme kinetics.* W. B. Saunders, Philadelphia.

Corella, J., Frances, E. 1991. Fluid catalytic cracking II. *ACS Symposium Series* **452**:165–182.

De Donder, T. 1920. *Leçons de thermodynamique et de chimie-physique.* Gauthier-Villus, Paris.

Drapper, N.R., Smith, H. 1981. *Applied regression analysis.* John Wiley & Sons, Inc., New York.

Eadie, G.S. 1942. The inhibition of cholinesterase by physostigmine and prostigmine. *J. Biol. Chem.* **146**:85–93.

Feng, W., Vynckier, E., Froment, G.F. 1993. Single event kinetics of catalytic cracking. *Ind. Eng. Chem. Res.* **32**:2997–3005.

Fogler, H.S. 1992. *Elements of chemical reaction engineering,* 2nd ed. Prentice-Hall, Englewood Cliffs, NJ.

Freitez, J., Peraza, A., Vargas, R., Verruschi, E. 2005. *Estudio sobre la estimación de parámetros cinéticos en el proceso de hidrodesulfuración del gasóleo.* Cong. Interam. Ing. Quim., Lima, Perú.

Froment, G.F., Bischoff, K.B., De Wilde, J. 2010. *Chemical reactor analysis and design,* 3rd ed. John Wiley & Sons, Inc., New York.

Hari, C., Balaraman, K.S., Balakrishnan, A.R. 1995. Fluid catalytic cracking: selectivity and product yield patterns. *Chem. Eng. Technol.* **18**:364–369.

Helfferich, F.G. 2003. *Kinetics of homogeneous multistep reactions,* 2nd ed. Elsevier Science, Amsterdam.

Henri, V. 1902. Théorie générale de l'action de quelques diastases. *C. R. Acad. Sci. Paris* **135**:916–919.

Hill, C.G. 1977. *An introduction to chemical engineering kinetics and reactor design.* John Wiley & Sons, Inc., New York.

Himmelblau, D.M. 1970. *Process analysis by statistical methods.* John Wiley & Sons, Inc., New York.

Hinshelwood, C.H., Askey, P.S. 1927. Homogeneous reactions involving complex molecules: the kinetics of the decomposition of gaseous dimethyl ether. *Proc. Roy. Soc.* **A115**:215–226.

Holland, C.D., Anthony, R.G. 1979. *Fundamentals of chemical reaction engineering.* Prentice-Hall, Englewood Cliffs, NJ.

Hu, M., Shao, H. 2006. Theory analysis of nonlinear data reconciliation and application to a coking plant. *Ind. Eng. Chem. Res.* **45**:8973–8984.

Kilanowski, D.R., Gates, B.C. 1980. Kinetics of hydrodesulfurization of benzothiophene catalyzed by sulfided Co-Mo/Al$_2$O$_3$. *J. Catal.* **62**:70–78.

Kistrakowsky, G.B., Lacker, J.R. 1936. The kinetics of some gaseous diels-alder reactions. *J. Am. Chem. Soc.* **58**:123–133.

Krambeck, F.J. 1991. *An industrial viewpoint on lumping: kinetics and thermodynamic lumping of multicomponent mixtures.* Elsevier Science, Amsterdam.

Lee, L.S., Chen, Y.W., Huang, T.N., Pan, W.Y. 1989. Four-lump kinetic model for fluid catalytic cracking process. *Can. J. Chem. Eng.* **67**:615–619.

Levenspiel, O. 1972. *Chemical reaction engineering*, 2nd ed. John Wiley & Sons, Inc., New York.

Levenspiel, O. 1979. *The chemical reactor omnibook.* Oregon State University Press. Corvallis.

Lineweaver, H., Burk, D. 1934. The determination of enzyme dissociation constants. *J. Am. Chem. Soc.* **56**:658–666.

Maronna, R., Arcas, J. 2009. Data reconciliation and gross error diagnosis based on regression. *Comp. Chem. Eng.* **33**:65–71.

Marquardt, D.W. 1963. An algorithm for least-squares estimation of nonlinear parameters. *J. Soc. Ind. Appl. Math.* **11**:431–441.

Matsumura, A., Kondo, T., Sato, S., Saito, I., De Souza, W. 2005. Hydrocracking Brazilian Marlim vacuum residue with natural limonite. Part 1: catalytic activity of natural limonite. *Fuel.* **84**:411–416.

Miao, Y., Su, H., Xu, O., Chu, J. 2009. Support vector regression approach for simultaneous data reconciliation and gross error or outlier detection. *Ind. Eng. Chem. Res.* **48**:10903–10911.

Michaelis, L., Menten, M.L. 1913. *Die Kinetik der Invertinwirkung. Biochem Z.* **49**:333–369.

Nguyen, T.S., Tayakout-Fayolle, M., Ropars, M., Geantet, C. 2013. Hydroconversion of an atmospheric residue with a dispersed catalyst in a batch reactor: kinetic modeling including vapor-liquid equilibrium. *Chem. Eng. Sci.* **94**:214–223.

Oliveira, C.E., Aguiar, P.F. 2009. Data reconciliation in the natural gas industry: analytical applications. *Energy Fuels.* **23**:3658–3664.

Perego, C., Peratello, S. 1999. Experimental methods in catalytic kinetics. *Catal. Today.* **52**:133–145.

Phillips, G.A., Harrison, D.P. 1993. Gross error detection and data reconciliation in experimental kinetics. *Ind. Eng. Chem. Res.* **32**:2530–2536.

Pinheiro, I.C.C., Fernandes, J.L., Domingues, L., Chambeln, A.J.S., Gracia, I., Oliveira, N.M.C., Cerqueira, H.S., Ribeiro, F.R. 2012. Fluid catalytic cracking (FCC) process modeling. *Ind. Eng. Chem. Res.* **51**:1–29.

Pitault, I. Fongarland, P., Mitrovic, M. 2004. Choice of laboratory scale reactors for HDT kinetics studies or catalyst tests. *Catal. Today.* **98**:31–42.

Quastel, J.H., Woolf, B. 1926. The equilibrium between *l*-aspartic acid, fumaric acid and ammonia in presence of resting bacteria. *Biochem. J.* **20**:545–555.

Reklaitis, G.V., Ravindran, A., Ragsdell, K. 1983. *Engineering optimization methods and applications.* John Wiley & Sons, Inc., New York.

Rezaei, H., Liu, X., Ardakani, S.J., Smith, K.J., Bricker, M. 2010. A study of Cold Lake vacuum residue hydroconversion in batch and semi-batch reactors using dispersed catalysts. *Catal. Today.* **150**:244–254.

Rice, N. 1997. The temperature scanning reactor II: theory of operation. *Catal. Today.* **36**:191–207.

Rubinstein, R. 1981. *Simulation and the Monte Carlo method.* John Wiley & Sons, Inc., New York.

Sadeghbeigi, R. 1995. *Fluid catalytic cracking handbook.* Houston, TX: Gulf Publishing.

Seferlis, P., Hrymak, A.N. 1996. Sensitivity analysis for chemical process optimization. *Comp. Chem. Eng.* **20**:1177–1200.

Smaïli, F., Vassiliadis, V.S., Wilson, D.I. 2001. Mitigation of fouling in refinery heat exchanger networks by optimal management of cleaning. *Energy Fuels.* **15**:1038–1056.

Smith, J.M., Van Ness, H.C., Abbott, M.M. 1980. *Introduction to chemical engineering thermodynamics.* McGraw-Hill, New York.

Tong, H., Crowe, C.M. 1995. Detection of gross errors in data reconciliation by principal component analysis. *AIChE J.* **41**:1712–1722.

Tukey, J.W. 1977. *Exploratory data analysis.* Addison-Wesley, Reading, MA.

Van Landeghem, F., Nevicato, D., Pitault, I., Forissier, M., Turlier, P., Derouin, C., Bernard, J.R. 1996. Fluid catalytic cracking: modelling of an industrial riser. *Appl. Cat. A.* **138**:381–405.

Varma, A., Morbidelli, M., Wu, H. 1999. *Parametric sensitivity in chemical systems.* Cambridge University Press, Cambridge.

Vasebi, A., Poulin, E., Hodouin, D. 2011. Observers for mass and energy balance calculation in metallurgical plants. Proceedings of the 18th IFAC World Congress, Milan, Italy, 9935–9940.

Wallenstein, D., Alkemade, U. 1996. Modelling of selectivity data obtained from microactivity testing of FCC catalysts. *Appl. Catal. A* **137**:37–54.

Wang, D., Romagnoli, J.A. 2003. A framework for robust data reconciliation based on a generalized objective function. *Ind. Eng. Chem. Res.* **42**:3075–3084.

Wang, F., Jia, S., Zheng, X., Yue, J. 2004. An improved MT–NT method for gross error detection and data reconciliation. *Comp. Chem. Eng.* **28**: 2189–2192.

Wang, Y. 1970. Catalytic cracking of gas oils. PhD dissertation, Chemical and Fuels Engineering Department, University of Utah.

Weekman, V.W. 1968. A model of catalytic cracking conversion in fixed, moving and fluid-bed reactors. *Ind. Eng. Chem. Proc. Des. Dev.* **7**:90–95.

Weekman, V.W. 1969. Kinetics and dynamics of catalytic cracking selectivity in fixed beds. *Ind. Eng. Chem. Proc. Des. Dev.* **8**:385–391.

Wen Li, H., Andersen, T.R., Gani, R., Jørgensen, S.B. 2006. Operating pressure sensitivity of distillations control structure consequences. *Ind. Eng. Chem. Res.* **45**:8310–8318.

Wojciechowski, B. 1997. The temperature scanning reactor I: reactor types and modes of operation. *Catal. Today.* **36**:167–190.

Worsfold, D.J., Bywater, S. 1960. Anionic polymerization of styrene. *Can. J. Chem.* **38**(10):1891–1900.

Wynkoop, R., Wilhelm, R.H. 1950. Kinetics in tubular reactor, hydrogenation of ethylene over copper-magnesia catalyst. *Chem. Eng. Progr.* **46**:300–310.

Zhang, Z., Shao, Z., Chen, X., Wang, K., Qian, J. 2010. Quasi-weighted least squares estimator for data reconciliation. *Comp. Chem. Eng.* **34**:154–162.

Index

Page numbers in *italics* refer to illustrations; page numbers in **bold** indicate a table.